T0200716

Symmetry and Its Discontents

This volume brings together a collection of essays on the history and philosophy of probability and statistics by one of the eminent scholars in these subjects. Written over the last fifteen years, they fall into three broad categories. The first deals with the use of symmetry arguments in inductive probability, in particular, their use in deriving rules of succession (Carnap's "continuum of inductive methods"). The second group deals with four outstanding individuals who made lasting contributions to probability and statistics in very different ways: Frank Ramsey, R. A. Fisher, Alan Turing, and Abraham de Moivre. The last group of essays deals with the problem of "predicting the unpredictable" – making predictions when the range of possible outcomes is unknown in advance. The essays weave together the history and philosophy of these subjects and document the fascination that they have exercised for more than three centuries.

S. L. Zabell is professor of mathematics and statistics at Northwestern University. A Fellow of the Institute of Mathematical Statistics and the American Statistical Association, he serves on the editorial boards of Cambridge Studies in Probability, Induction, and Decision Theory, and The Collected Works of Rudolph Carnap. He received the Distinguished Teaching Award from Northwestern University in 1992.

**Cambridge Studies in Probability,
Induction, and Decision Theory**

General editor: Brian Skyrms

Advisory editors: Ernest W. Adams, Ken Binmore, Jeremy Butterfield,
Persi Diaconis, William L. Harper, John Harsanyi,
Richard C. Jeffrey, James M. Joyce, Wlodek Rabinowicz,
Wolfgang Spohn, Patrick Suppes, Sandy Zabell

Symmetry and Its Discontents

Essays on the History of Inductive Probability

S. L. ZABELL

Northwestern University

CAMBRIDGE
UNIVERSITY PRESS

CAMBRIDGE UNIVERSITY PRESS
Cambridge, New York, Melbourne, Madrid, Cape Town, Singapore, São Paulo

Cambridge University Press
40 West 20th Street, New York, NY 10011-4211, USA

www.cambridge.org
Information on this title: www.cambridge.org/9780521444705

First published 2005

Printed in the United States of America

A catalog record for this publication is available from the British Library.

Library of Congress Cataloging in Publication Data
Zabell, S. L., 1947–
Symmetry and its discontents : essays on the history of inductive probability / S. L. Zabell.
p.   cm. – (Cambridge studies in probability, induction, and decision theory)
Includes bibliographical references and index.
ISBN 0-521-44470-5 – ISBN 0-521-44912-X (pbk.)
1. Probabilities – History.   2. Induction (Logic).   3. Inference.   4. Prediction (Logic).
5. Mathematicians – History.   I. Title.   II. Series.
QA273.18Z33   2006
519.2′09 – dc22   2005006325

ISBN-13   978-0-521-44470-5 hardback
ISBN-10   0-521-44470-5 hardback

ISBN-13   978-0-521-44912-0 paperback
ISBN-10   0-521-44912-X paperback

For Dick Jeffrey, mentor and friend.

Contents

Preface

These essays tell the story of inductive probability, from its inception in the work of Thomas Bayes to some surprising current developments. Hume and Bayes initiated a dialogue between inductive skepticism and probability theory that persists in various forms throughout the history of the subject. The skeptic insists that we start in a state of ignorance. How does one quantify ignorance? If knowledge gives rise to asymmetric probabilities, perhaps ignorance is properly characterized by symmetry. And non-trivial prior symmetries generate non-trivial inductive inference. Then perhaps symmetries are not quite such innocent representations of ignorance as one might have thought. That is a sketch of the theme that is developed in the title essay, "Symmetry and its Discontents", and that runs throughout the book.

In the second section of this book, we meet Sir Alexander Cuming, who instigated important investigations by De Moivre and Stirling, before being sent to prison for fraud. We view Ramsey's famous essay "Truth and Probability" against the Cambridge background of Robert Leslie Ellis, John Venn and John Maynard Keynes. Fisher's discussion of inverse probabilities is set in the context of Boole, Venn, Edgeworth and Pearson and his various versions of the fiducial argument are examined. We learn of Alan Turing's undergraduate rediscovery of Lindeberg's central limit theorem, and of his later use of Bayesian methods in breaking the German naval code in World War II.

The last section deals with developments in inductive probability, which are still not generally well-known, and that some philosophers have thought impossible. The question is how a Bayesian theory can deal in a principled way with the possibility of new categories that have not been foreseen. On the face of it the problem appears to be intractable, but a deeper analysis shows that something sensible can be done. The development of the appropriate mathematics is a story that stretches from the beginnings of the subject to the end of the twentieth century.

These essays have appeared, over almost twenty years, in a variety of disparate and sometimes obscure places. I remember eagerly waiting for the next installment. Each essay is like a specially cut gem, and it gives me great satisfaction that they can be brought together and presented in this volume.

Brian Skyrms

PART ONE

Probability

1

Symmetry and Its Discontents

The following paper consists of two parts. In the first it is argued that Bruno de Finetti's theory of subjective probability provides a partial resolution of Hume's problem of induction, if that problem is cast in a certain way. De Finetti's solution depends in a crucial way, however, on a symmetry assumption – exchangeability – and in the second half of the paper the broader question of the use of symmetry arguments in probability is analyzed. The problems and difficulties that can arise are explicated through historical examples which illustrate how symmetry arguments have played an important role in probability theory throughout its development. In a concluding section the proper role of such arguments is discussed.

1. THE DE FINETTI REPRESENTATION THEOREM

Let $X_1, X_2, X_3, \ldots$ be an infinite sequence of 0,1-valued random variables, which may be thought of as recording when an event occurs in a sequence of repeated trials (e.g., tossing a coin, with 1 if heads, 0 if tails). The sequence is said to be *exchangeable* if all finite sequences of the same length with the same number of ones have the same probability, i.e., if for all positive integers n and permutations σ of $\{1, 2, 3, \ldots, n\}$,

$$P[X_1 = e_1, X_2 = e_2, \ldots, X_n = e_n]$$
$$= P[X_1 = e_{\sigma(1)}, X_2 = e_{\sigma(2)}, \ldots, X_n = e_{\sigma(n)}],$$

where e_i denotes either a 0 or a 1. For example, when $n = 3$, this means that

$$P[1, 0, 0] = P[0, 1, 0] = P[0, 0, 1] \quad \text{and}$$
$$P[1, 1, 0] = P[1, 0, 1] = P[0, 1, 1].$$

(Note, however, that $P[1, 0, 0]$ is not assumed to equal $P[1, 1, 0]$; in general, these probabilities may be quite different.)

Reprinted with permission from Brian Skyrms and William L. Harper (eds.), *Causation, Chance, and Credence* 1 (1988): 155–190, © 1988 by Kluwer Academic Publishers.

In 1931 the Italian probabilist Bruno de Finetti proved his famous *de Finetti Representation Theorem. Let $X_1, X_2, X_3, \ldots$ be an infinite exchangeable sequence of $0,1$-valued random variables, and let $S_n = X_1 + X_2 + \cdots + X_n$ denote the number of ones in a sequence of length n. Then it follows that:*

1. *the limiting frequency $Z =: \lim_{n \to \infty}(S_n/n)$ exists with probability 1.*
2. *if $\mu(A) =: P[Z \in A]$ is the probability distribution of Z, then*

$$P[S_n = k] = \int_0^1 \binom{n}{k} p^k (1-p)^{n-k} \mathrm{d}\mu(p)$$

for all n and k.[1]

This remarkable result has several important implications. First, contrary to popular belief, subjectivists clearly believe in the existence of infinite limiting relative frequencies – at least to the extent that they are willing to talk about an (admittedly hypothetical) infinite sequence of trials.[2] The existence of such limiting frequencies follows as a purely mathematical consequence of the assumption of exchangeability.[3] When an extreme subjectivist such as de Finetti denies the existence of objective chance or physical probability, what is really being disputed is whether limiting frequencies are objective or physical properties.

There are several grounds for such a position, but all center around the question of what "object" an objective probability is a property of. Surely not the infinite sequence, for that is merely a convenient fiction (Jeffrey 1977). Currently the most fashionable stance seems to be that objective probabilities are a *dispositional property* or *propensity* which manifests itself in, and may be measured with ever-increasing accuracy by, finite sequences of ever-increasing length (e.g., Kyburg 1974).

But again, a property of what? Not the coin, inasmuch as some people can toss a so-called fair coin so that it lands heads 60% of the time or even more (provided the coin lands on a soft surface such as sand rather than a hard surface where it can bounce). Some philosophers attempt to evade this type of difficulty by ascribing propensities to a *chance set-up* (e.g., Hacking 1965): in the case of coin-tossing, the coin *and* the manner in which it is tossed. But if the coin were indeed tossed in an identical manner on every trial, it would always come up heads or always come up tails; it is precisely because the manner in which the coin is tossed on each trial is *not* identical that the coin can come up both ways. The suggested chance set-up is in fact nothing other than a sequence of objectively differing trials which we are subjectively unable to distinguish between. At best, the infinite limiting frequency is a property of an "object" enjoying both objective and subjective features.

2. DE FINETTI VANQUISHES HUME

The most important philosophical consequence of the de Finetti representation theorem is that it leads to a solution to *Hume's problem of induction:* why should one expect the future to resemble the past? In the coin-tossing situation, this reduces to: in a long sequence of tosses, if a coin comes up heads with a certain frequency, why are we justified in believing that in future tosses of the same coin, it will again come up heads (approximately) the same fraction of the time?

De Finetti's answer to this question is remarkably simple. Given the information that in n tosses a coin came up heads k times, such data is incorporated into one's probability function via

Bayes's rule of conditioning: $P[A \mid B] = P[A \text{ and } B]/P[B]$.

If n is large and $p^* = k/n$, then – except for certain highly opinionated, eccentric, or downright kinky "priors" $d\mu$ – it is easy to prove that the resulting posterior probability distribution on p will be highly peaked about p^*; that is, the resulting probability distribution for the sequence of coin tosses looks approximately like (in a sense that can be made mathematically precise) a sequence of independent and identically distributed Bernoulli trials with parameter p^* (i.e., independent tosses of a p^* coin). By the weak law of large numbers it follows that, with high probability, subsequent tosses of the coin will result in a relative frequency of heads very close to p^*.

Let us critically examine this argument. Mathematically it is, of course, unassailable. It implicitly contains, however, several key suppositions:

1. P is operationally defined in terms of betting odds.
2. P satisfies the axioms of mathematical probability.
3. P is modified upon the receipt of new information by Bayesian conditioning.
4. P is assumed to be exchangeable.

In de Finetti's system, degree of belief is quantified by the betting odds one assigns to an event. By a Dutch book or coherence argument, one deduces that these betting odds should be consistent with the axioms of mathematical probability. Conditional probabilities are initially defined in terms of conditional bets and Bayes's rule of conditioning is deduced as a consequence of coherence. The relevance of conditional probabilities to inductive inference

5

is the *dynamic assumption of Bayesianism* (Hacking 1967): if one learns that B has occurred, then one's new probability assignment is $P[A \mid B]$. In general, however, conditional probabilities can behave in very nonHumeian ways, and (infinite) exchangeability is taken as describing the special class of situations in which Humeian induction is appropriate.

This paper will largely concern itself with the validity of this last assumption. Suffice it to say that, like Ramsey (1926), one may view the subjectivist interpretation as simply capturing *one* of the many possible meanings or useages of probability; that the Dutch book and other derivations of the axioms may be regarded as plausibility arguments (rather than normatively compelling); and that although a substantial literature has emerged in recent decades concerning the limitations of Bayesian conditioning, the difficulties discussed and limitations raised in that literature do not seem particularly applicable to most of the situations typically envisaged in discussions of Hume's problem.

The assumption of exchangeability, however, seems more immediately vulnerable. Isn't it essentially circular, in effect assuming what one wishes to prove? Of course, in one sense this must obviously be the case. All mathematics is essentially tautologous, and any implication is contained in its premises. Nevertheless, mathematics has its uses. Formal logic and subjective probability are both theories of consistency, enabling us to translate certain assumptions into others more readily palatable.

What de Finetti's argument really comes down to is this: if future outcomes are viewed as exchangeable, i.e., no one pattern is viewed as any more or less likely than any other (with the same number of successes), then when an event occurs with a certain frequency in an initial segment of the future, we must, if we are to be consistent, think it likely that that event will occur with approximately the same frequency in later trials. Conversely, if we do not accept this, it means that we must have – prospectively – thought certain patterns more likely than others. Which means that we must have possessed more information than is ordinarily posited in discussions of Humeian induction.

And there the matter would appear to stand. Or does it?

3. THE INSIDIOUS ASSUMPTION OF SYMMETRY

Exchangeability is one of many instances of the use of symmetry arguments to be found throughout the historical development of mathematical probability and inductive logic. But while such arguments often have a seductive attraction, they also often carry with them "hidden baggage": implications or consequences, sometimes far from obvious, which later cast serious doubt on their validity. We will discuss three historically important examples, all involving attempts to justify induction by the use of probability theory, and all (in effect) involving the appropriate choice of prior $d\mu$ in the de Finetti representation.

Example 3.1. *Bayes's argument for the Bayes–Laplace prior.*

Consider "an event concerning the probability of which we absolutely know nothing antecedently to any trials made concerning it" (Bayes 1764). Implicitly invoking a symmetry argument, Bayes argued that "concerning such an event I have no reason to think that, in a certain number of trials, it should rather happen any one possible number of times than another," i.e., that in a sequence of n trials one's probability assignment for S_n, the number of heads, should satisfy

Bayes's Postulate: $P[S_n = k] = 1/(n + 1)$.

That is, the number of heads can assume any of the $n + 1$ values $0, 1, 2, \ldots, n$ and, absent further information, all $n + 1$ values are viewed as equally likely. In a famous Scholium, Bayes concluded that if this were indeed the case, then the prior probability $d\mu(p)$ must be the "flat" prior dp.[4]

Although Bayes's exact reasoning at this point is somewhat unclear, it can easily be made rigorous: Taking $k = n$ in the de Finetti representation and using Bayes's postulate, it follows that

$$\int_0^1 p^n d\mu(p) = 1/(n + 1).$$

The integral on the left-hand side is the n-th moment of $d\mu$, so Bayes's assumption uniquely determines the moments of $d\mu$. But since $d\mu$ is concentrated on a compact set, it follows by a theorem of Hausdorff that $d\mu$, if it exists, is in turn determined by its moments. That is, there can be at most one probability measure $d\mu$ which satisfies Bayes's assumption $P[S_n = k] = 1/(n + 1)$. But the flat measure dp does satisfy this integral equation, i.e.,

$$\int_0^1 p^n dp = 1/(n + 1),$$

hence $d\mu$ must be dp.

Bayes's argument is quite attractive. A modern-day subjectivist might view Bayes's assumption as a *definition* (possibly one of many) of "complete ignorance" (rather than consider "complete ignorance" to be an *a priori* meaningful concept), but would probably find Bayes's argument otherwise unobjectionable.

The argument in its original form, however, did not go uncriticized. As Boole (1854, pp. 369–375) noted, rather than consider the events $[S_n = k]$ to be equally likely, one could equally plausibly take all sequences of a fixed length (or "constitutions") to be so. Thus, for $n = 3$

$$P[000] = P[100] = P[010] = P[001] = P[110]$$
$$= P[101] = P[011] = P[111] = 1/8.$$

To many, this assignment seemed a far more natural way of quantifying ignorance than Bayes's.

Unfortunately, it contains a time-bomb with a very short fuse. As Carnap (1950, p. 565) later noted (and Boole himself had already remarked), this probability assignment corresponds to *independent* trials, and thus remains unchanged when conditioned on the past, an obviously unsatisfactory choice for modeling inductive inference, inasmuch as "past experience does not in this case affect future expectation" (Boole 1854, p. 372).

In his *Logical Foundations of Probability* (1950), Carnap announced that in a later volume, "a quantitative system of inductive logic" would be constructed, based upon a function Carnap denoted c^*. Carnap's c^* function was, in effect, the one already proposed by Bayes. But Carnap grew uneasy with this unique choice, and in his monograph *The Continuum of Inductive Methods* (1952), he advocated instead the use of a one-parameter family containing c^*. Unknown to Carnap, however, he had been anticipated in this, almost a quarter of a century earlier, by the English philosopher William Ernest Johnson.

Example 3.2. *W. E. Johnson's sufficientness postulate.*

In 1924 Johnson, a Cambridge logician, proposed a multinomial generalization of Bayes's postulate. Suppose there are $t \geq 2$ categories or types, and in n trials there are n_1 outcomes of the first type, n_2 outcomes of the second type, ..., and n_t outcomes of the t-th type, so that $n = n_1 + n_2 + \cdots + n_t$. The sequence $(n_1, n_2, \ldots, n_t)$ is termed an *ordered t-partition of n*. Bayes had considered the case $t = 2$, and his postulate is equivalent to assuming that all ordered 2-partitions $(k, n - k)$ are equally likely. Now Johnson proposed as its generalization

Johnson's combination postulate: Every ordered *t*-partition of *n* is equally likely.

For example, if $t = 3$ and $n = 4$, then there are 15 possible ordered 3-partitions of 4, viz.:

n_1	n_2	n_3
4	0	0
3	1	0
3	0	1
2	2	0
2	1	1
2	0	2
1	3	0
1	2	1
1	1	2
1	0	3
0	4	0
0	3	1
0	2	2
0	1	3
0	0	4

and each of these is assumed to be equally likely.

Johnson did not work with integral representations but, like Carnap, with finite sequences. In so doing he introduced a second postulate, his "permutation postulate." This was none other than the assumption of exchangeability, thus anticipating de Finetti (1931) by almost a decade! (If one labels the types or categories with the letters of a t-letter alphabet, exchangeability here means that all words of the same length, containing the same number of letters of each type, are equally likely). Together, the combination and permutation postulates uniquely determine the probability of any specific finite sequence. For example, if one considers the fifth partition in the table above, $4 = 2 + 1 + 1$, then there are twelve sequences which give rise to such a partition, viz.

x_1	x_2	x_3	x_4
1	1	2	3
1	1	3	2
1	2	1	3
1	2	3	1
1	3	1	2
1	3	2	1
2	1	1	3
2	1	3	1
2	3	1	1
3	1	1	2
3	1	2	1
3	2	1	1

and each of these are thus assumed to have probability $(1/15)(1/12) = 1/180$. The resulting probability assignment on finite sequences is identical with Carnap's c^*.

Despite its mathematical elegance, Johnson's "combination postulate" is obviously arbitrary, and Johnson was later led to substitute for it another, more plausible one, his "sufficientness postulate." This new postulate assumes for all n

Johnson's sufficientness postulate:

$$P[X_{n+1} = j|X_1 = i_1, X_2 = i_2, \ldots, X_n = i_n] = f(n_j, n).$$

That is, the conditional probability that the next outcome is of type j depends only on the number of previous trials and the number of previous outcomes of type j, but not on the frequencies of the other types or the specific trials on which they occurred. If, for example $t = 3$, $n = 10$, and $n_1 = 4$, the postulate asserts that on trial 11 the (conditional) probability of obtaining a 1 is the same for all sequences containing four 1's and 6 not –1's, and that this conditional probability does not depend on whether there were six 2's and no 3's, or five 2's and one 3, and so on. (Note that the postulate implicitly assumes that all finite sequences have positive probability, so that the conditional probabilities are well-defined.)

Johnson's sufficientness postulate makes what seems a minimal assumption: absence of knowledge about different types is interpreted to mean that information about the frequency of one type conveys no information about the likelihood of other types occurring. It is therefore rather surprising that it follows from the postulate that the probability function P is uniquely determined up to a constant:

Theorem (Johnson 1932). *If P satisfies the sufficientness postulate and $t \geq 3$, then either the outcomes are independent or there exists a $k > 0$ such that*

$$f(n_i, n) = \{n_i + k\}/\{n + tk\}.$$

This is, of course, nothing other than Carnap's "continuum of inductive methods."[5]

The de Finetti representation theorem can be generalized to a much wider class of infinite sequences of random variables than those taking on just two values (e.g., Hewitt and Savage 1955). In the multinomial case now being discussed, the de Finetti representation states that every exchangeable probability can be written as a mixture of multinomial probabilities. Just as Bayes's postulate implied that the prior $d\mu$ in the de Finetti representation was the flat prior, Johnson's theorem implies that the mixing measure $d\mu$ in

10

the de Finetti representation is the *symmetric Dirichlet prior*

$$\Gamma(tk)/\Gamma(k)^t \, p_1^{k-1} p_2^{k-1} \ldots p_1^{k-1} \mathrm{d}p_1 \mathrm{d}p_2 \ldots \mathrm{d}p_{t-1}:$$

a truly remarkable result, providing a subjectivistic justification for the use of the mathematically attractive Dirichlet prior.[6]

Despite its surface plausibility, Johnson's sufficientness postulate is often too strong an assumption. While engaged in cryptanalytic work for the British government at Bletchley Park during World War II, the English logician Alan Turing realized that even if one lacks specific knowledge about individual category types, the frequencies $n_1, n_2, \ldots, n_t$ may contain relevant information about predictive probabilities, namely the information contained in the *frequencies of the frequencies*.

Let $a_r =$ the number of frequencies n_i equal to r; a_r is called the frequency of the frequency r. For example, if $t = 4$, $n = 10$, and one observes the sequence 4241121442, then $n_1 = 3, n_2 = 3, n_3 = 0, n_4 = 4$ and $a_0 = 1, a_1 = 0, a_2 = 0, a_3 = 2, a_4 = 1$. (A convenient shorthand for this is $0^1 1^0 2^0 3^2 4^1$.) Although it is far from obvious, the a_r may be used to estimate cell probabilities: see Good (1965, p. 68).[7]

Example 3.3. *Exchangeability and partial exchangeability.*

Given the failure of such attempts, de Finetti's program must be seen as a further retreat from the program of attempting to provide a unique, quantitative account of induction. Just as Johnson's sufficientness postulate broadened the class of inductive probabilities from that generated by the Bayes–Laplace prior to the continuum generated by the symmetric Dirichlet priors, so de Finetti extended the class of possible inductive probabilities even further to include *any* exchangeable probability assignment.

But what of the symmetry assumption of exchangeability? Even this is not immune to criticism (as de Finetti himself recognized). Consider the following sequence: 000101001010100010101001.... Scrutiny of the sequence reveals the interesting feature that although every 0 is followed by a 0 or 1, every 1 is invariably followed by a 0. If this feature were observed to persist over a long segment of the sequence (or simply that 1's were followed by 0's with high frequency), then this would seem relevant information that should be taken into account when calculating conditional, predictive probabilities. Unfortunately, exchangeable probabilities are useless for such purposes: if P is exchangeable, then the conditional probabilities

$$P[X_{n+1} = j | X_1 = i_1, X_2 = i_2, \ldots, X_n = i_n]$$

depend solely on the number of 1's, and not on their order within the sequence. Thus, exchangeability, despite its plausibility, rules out a natural form of inductive inference and can only be considered valid when "order effects" are ruled out (as, for example, in coin-tossing).

An appropriate generalization of exchangeability that takes such order information into account is the concept of *Markov exchangeability:* all sequences with the same initial letter and the same transition counts ($t_{ij} =:$ number of transitions from state i to state j) are assumed equally likely. Here too a de Finetti representation is possible (Diaconis and Freedman 1980b, 1980c): now one mixes on the possible transition matrices p_{ij}.

Once one has come this far, of course, it is easy to recognize that order effects of this type are merely one of many possible patterns that may be judged to provide useful information, each pattern requiring a corresponding generalization of exchangeability to incorporate the information it provides. To deal with such situations, de Finetti introduced in 1938 the notion of *partial exchangeability* (Diaconis and Freedman 1980c). Although partial exchangeability is an active field of current mathematical research still undergoing development (e.g., Diaconis and Freedman 1985), the general outline of the theory is clear: to each pattern corresponds a statistic or symmetry, a representation theorem, and a corresponding mode of inductive inference.

Thus, de Finetti's resolution of Hume's problem of induction is a highly qualified one: it is a theory of coherence. Every person's probability function will contain some symmetry involving past and future, and coherence dictates that patterns observed in the past will be expected to recur in the future.

Despite its highly qualified nature, the above analysis has an important payoff: it demonstrates that Hume's problem is in fact illposed; to ask "*why* should the future be expected to resemble the past?" presupposes having already answered the question "*how* is the future expected to resemble the past?" (It is essentially this point that is behind Nelson Goodman's "grue" paradox.) It is a strength of the subjectivist analysis that this point emerges as natural and obvious; indeed, it is essentially forced on one; and to the extent that one can state precisely the ways in which the past and future are conjectured to correspond, it gives a satisfactory solution to Hume's problem.

The successive attempts of Bayes, Johnson, and de Finetti to solve the problem of induction are marked by the invocation of progressively weaker symmetry assumptions. Symmetry, however, has played not only a key role in the attempts to quantify induction, it has played a central role in the birth and evolution of probability theory, more central perhaps than sometimes recognized. In the next three sections it will be argued that the birth of mathematical probability marked a key change in the way symmetry arguments were used; that the early dependence on symmetry arguments to quantify probability, while crucial to its mathematical development, blurred important epistemological distinctions; and that it was only with the challenging of precisely those symmetry arguments in the nineteenth century that the conceptual clarification of probability became possible.

4. OU MALLON

The simplest and oldest of such arguments is the use of physical or epistemic symmetry to identify a *fundamental probability set* or FPS, i.e., a partition of the space of possible outcomes into equiprobable alternatives. The recognition and use of such sets to compute numerical probabilities for complex events was a key step in the birth of mathematical probability. Once the ability to calculate probabilities in this simple case had been mastered, the outlines of the mathematical theory discerned, and its practical utility recognized, all else followed. Why were the mathematicians of the seventeenth century able to take this step, while the Greeks, despite their mathematical prowess and penchant for games of chance, were not? The crucial point to recognize is that while for the pioneers of the modern theory the equipossible elements of an FPS were *equally likely*, for the Greeks *none were possible*.

This was because of what G. E. L. Owen has described as "a very Greek form of argument" (Owen 1966), a form of reasoning employed by the Greeks that Leibniz was very fond of and which he called the *principle of sufficient reason*: "for every contingent fact there is a reason why the fact is so and not otherwise . . . " (Broad 1975, p. 11). In the words of Leucippus (the only complete sentence of his which has come down to us), "Nothing occurs at random, but everything for a reason and by necessity" (Kirk and Raven 1957, p. 413). Two famous examples will illustrate its use:

4.1. Anaximander and the Position of the Earth

Anaximander (c. 610–540 B.C.), one of the early pre-Socratic Greek philosophers, believed the Earth lay at the center of the universe. But unlike Thales before him, who thought the Earth floated on water, and Anaximenes after, who thought it floated on air, Anaximander thought the Earth was unsupported and remained at the center for reasons of symmetry (*omoiotes*; variously translated as "similarity," "indifference," "equilibrium," or "equiformity").[8] Unfortunately, the text of Anaximander has not survived, and we are dependent on secondary, incomplete, and contradictory later accounts for information about the precise nature of his astronomical beliefs.[9] Our best source is perhaps Aristotle, who reports:

There are some who say, like Anaximander among the ancients, that [the earth] stays still because of its equilibrium. For it behoves that which is established at the center, and is equally related to the extremes, not to be borne one whit more either up or down or to the sides; and it is impossible for it to move simultaneously in opposite directions, so that it stays fixed by necessity. [*de Caelo* 295 b10]

How closely this reproduces Anaximander's own logic, the exact meaning to be attached to *omoiotes*, indeed the precise nature of the argument itself, is unclear. Nevertheless, the gist of the argument is clearly an appeal to symmetry: for every direction there is an opposite; since there is no more reason for the earth to move in one direction than another, the proper conclusion is that it moves in *neither*.

Although Aristotle expressed scepticism about such reasoning, it was fully accepted by Plato:

I am therefore persuaded that, in the first place, since the earth is round and in the middle of the heaven, it has not need either of air or any other necessity in order not to fall, but the similarity of the heaven to itself in every way and the equilibrium of the earth suffice to hold it still. For an equilibrated thing set in the midst of something of the same kind will have no reason to incline in one direction more than in another. But as its relationship is symmetrical it will remain unswervingly at rest. [*Phaedo* 108e–109a; c.f. *Timaeus* 62d.12]

4.2. Parmenides and the Creation of the Universe

Parmenides gave a similar argument to show that the universe had never been created:

And what need would have driven it on to grow, starting from nothing, at a later time rather than an earlier? [Kirk and Raven 1957, p. 273]

Again this is essentially a symmetry argument: if the universe had been created, it must have been at some specific time; inasmuch as there is no more reason for it to have been created at any one time than any other, all possible times are thereby ruled out. Obviously the argument makes some presuppositions, but it had great appeal to Leibniz and appears in his correspondence with Clarke.[10]

It is, as G. E. L. Owen notes,

a very Greek pattern of argument. . . . Aristotle retailored the argument to rebut the probability of motion in a vacuum; the Academy adapted it to show that, since no physical sample of equality has more right to serve as a standard sample than any other, the standard sample cannot be physical. And Leibniz found an excellent example in Archimedes's mechanics. . . . [Owen 1966]

The Greek Pyrrhonian skeptics made systematic use of a similar device for destroying belief. Their goal was to achieve a state of *epoche*, or suspension of judgement about statements concerning the external world, which they believed would in turn lead to *ataraxia*, a state of tranquility, ". . . saying concerning each individual thing that it no more [*ou mallon*] is than is not, or that it both is and is not, or that it neither is nor is not."[11]

How can *epoche* be achieved? According to Sextus Empiricus (*Outlines of Pyrrhonism* 1.8):

Scepticism is an ability which sets up antitheses among appearances and judgments in any way whatever: by scepticism, on account of the 'equal weight' which characterizes opposing states of affairs and arguments, we arrive first at 'suspension of judgment', and second at 'freedom from disturbance'.

For example, knowledge of what is good is impossible, for what one person thinks good, another may think bad, and

if we say that not all that anyone thinks good is good, we shall have to judge the different opinions; and this is impossible because of the equal validity of opposing arguments. Therefore the good by nature is impossible.

It is important to understand the implications of asserting "*ou mallon*." One might interpret it in a positive sense: although certain knowledge is ruled out, the information we possess is equally distributed between two or more possibilities, and hence we have an equal degree of belief in each. That this was *not* the skeptical position is clear from a passage in Diogenes Laertius (*Life of Pyrrho* 9.74–76):

Thus by the expression "We determine nothing" is indicated their state of even balance; which is similarly indicated by the other expressions, "Not more (one thing than another)," "Every saying has its corresponding opposite," and the like. But "Not more (one thing than another)" can also be taken positively, indicating that two things are alike; for example, "The pirate is no more wicked than the liar." But the Sceptics meant it not positively but negatively, as when, in refuting an argument, one says, "Neither had more existence, Scylla or the Chimaera . . ." Thus, as Timon says in the *Pytho*, the statement [*ou mallon*] means just absence of all determination and withholding of assent. The other statement, "Every saying, etc.," equally compels suspension of judgment; *when facts disagree, but the contradictory statements have exactly the same weight, ignorance of the truth is the necessary consequence.* [Emphasis added]

Pyrrhonian skepticism is an extreme position, and the later Academic skeptics developed a theory that combined skepticism about certain knowledge with a description of rational decision based on probable knowledge.[12] Under Carneades this theory included a scale of the varying degrees of conviction conveyed by an impression, depending on whether it was "credible," "credible and consistent," or "credible, consistent, and tested." Carneades's theory amounts to an early account of qualitative or comparative subjective probability, and one might expect that a later skeptic would go the final step and attempt to numerically measure or describe such degrees of conviction. That this did not happen, it may be argued, was a consequence of the *ou mallon* viewpoint. Witness Cicero's statement:

If a question be put to [the wise man] about duty or about a number of other matters in which practice has made him an expert, he would not reply in the same way as he would if questioned as to whether the number of the stars is even or odd, and say that he did not know; for in things uncertain there is nothing probable [*in incertis enim nihil est probabile*], but in things where there is probability the wise man will not be at a loss either what to do or what to answer. [Cicero *Academica* 2.110]

A 19th century enthusiast of the principle of insufficient reason would have little hesitation in assigning equal probabilities to the parity of the number of stars; this passage thus strikingly illustrates a chasm that had to be crossed before numerical probabilities could be assigned. Cicero was familiar with a theory of probability, indeed much of the *Academica* is devoted to a discussion of Academic probabilism and is one of our major sources of information about it. But for Cicero the probable was limited in its scope, limited in a way that precluded its quantification. The FPS was the basic setting for the early development of mathematical probability – but for Cicero it was a setting in which the very notion of probability itself was inapplicable.

Support for this thesis may be found in the writings of Nicole Oresme, the Renaissance astronomer and mathematician (ca. 1325–1382). Oresme discussed Cicero's example of the number of stars but, writing only a few centuries before the earliest known probability calculations, there is a clear difference:

The number of stars is even; the number of stars is odd. One of these statements is necessary, the other impossible. However, we have doubts as to which is necessary,

16

so that we say of each that it is possible.... The number of stars is a cube. Now indeed, we say that it is possible but not, however, probable or credible or likely [*non tamen probabile aut opinabile aut verisimile*], since such numbers are much fewer than others.... The number of stars is not a cube. We say that it is possible, probable, and likely.... [Oresme 1966, p. 385]

To some, the revolutionary content of this passage lies in its quasinumerical assertion of the improbability of the number of stars being a cube (due to the infrequency of cubic numbers). But its real novelty is Oresme's willingness to extend the realm of the probable. Having made that transition, the frequency-based assertions of probability and improbability he makes follow naturally.

Thus the key step in the birth of mathematical probability – the identification of fundamental probability sets in order to quantify probability – while seemingly so natural, in fact contains a major presupposition. The ancients used symmetry arguments to destroy belief, where we use them to quantify it. This "conceptual revolution" culminated in the 20th century statistical resort to physical randomization (e.g., in sampling, randomized clinical trials, and Monte Carlo simulations): the paradox of deliberately imposing disorder to acquire information. The uses of randomization throughout the ancient and medieval world, in contrast, although common and widespread (for example, in games of chance and fair allocation) all depended, in one way or another, solely on its property of loss of information.

But while the use of symmetry made the calculus of probabilities possible, it also contained the seeds of future confusion.

5. CHANCE AND EQUIPOSSIBILITY

The birth of probability was not an untroubled one. Probabilities are usually classified into two major categories – epistemic and aleatory – and a multitude of subcategories: propensities, frequencies, credibilities, betting odds, and so on. In settings where an FPS exists, all of these will usually have a common value, and the necessity of distinguishing among the different meanings is not a pressing one. But as the initial successes of the "doctrine of chances" spurred on its application to other spheres, this happy state of affairs ceased and the need for distinctions became inevitable.

Just what the proper domains of chance and probability were, however, remained unclear. For the calculus of probabilities was initially the "doctrine of chances," and paradoxically, while the Greeks failed to extend the realm of the probable to include fundamental probability sets, in the early days of

17

the doctrine of chances some thought the notion of chance *only* applicable to such settings. A few examples will suggest the difficulties and confusions that occurred.

1. *Arbuthnot and the sex-ratio.* In 1711, Dr. John Arbuthnot, a Scottish writer, physician to Queen Anne, and close friend of Swift and Pope, published a short paper in the *Philosophical Transactions of the Royal Society*, titled, 'An Argument for Divine Providence Taken From the Constant Regularity Observed in the Births of Both Sexes.' Using statistics from the London Bills of Mortality for the preceding 82 years, Arbuthnot observed that male births had exceeded female births in London for each year from 1629 to 1710. Noting that if male and female births were equally likely, the probability of such an outcome was extremely small (1 in 2^{82}), Arbuthnot rejected the hypothesis of equilikelihood, making in effect the earliest known statistical test of significance. But Arbuthnot did not conclude that male and female births possessed unequal probabilities. Instead, he rejected outright the possibility that sex was due to chance, concluding that the excess of males was due to the intervention of divine providence; that "... it is Art, not Chance, that governs" (Arbuthnot 1711, p. 189).

In contrasting art with chance, Dr. Arbuthnot was merely displaying his classical erudition; the dichotomy between *techne* (art) and *tyche* (chance) being a commonplace of Greek philosophy.[13] What *is* new is his belief that chance is only operative when probabilities are equilikely; that otherwise some outside force must be acting, causing the imbalance, and that one could no longer refer to chance. His specific line of reasoning was quickly faulted by Nicholas Bernoulli: if sex is likened to tossing a 35-sided die, with 18 faces labelled "male," and 17 labelled "female," then Arbuthnot's data are entirely consistent with the outcome of chance.[14] This response to Arbuthnot's argument does not dispute that chance is limited to fundamental probability sets; it simply points out that more that one FPS is possible.

Arbuthnot's juxtaposition of chance and cause, and his belief that chances must be equal, is echoed in Hume. For Hume chance "properly speaking, is merely the negation of a cause":

Since therefore an entire indifference is essential to chance, no one chance can possibly be superior to another, otherwise than as it is compos'd of a superior number of equal chances. For if we affirm that one chance can, after any other manner, be superior to another, we must at the same time affirm, than there is something, which gives it superiority, and determines the event rather to that side than the other: That is, in other words, we must allow of a cause, and destroy the supposition of chance; which we had before establish'd. A perfect and total indifference is essential to chance, and one total indifference can never in itself be either superior or inferior to another. This truth is not peculiar to my system, but is acknowledg'd by every one, that forms calculations concerning chances. [*Hume* 1739, p. 125]

Thus, for Hume, not merely the mathematical calculation of chances but the very existence of chance itself is dependent on an "entire," "perfect," and "total indifference" among the different possibilities. Was this "acknowledg'd by every one?" Examination of the works of Bernoulli, DeMoivre, and Laplace does not entirely bear out this claim. There the equality of chances appears as a mathematical device, not a metaphysical necessity. Nevertheless, the contrast of chance with "art," "design," or "cause," that "something, which gives it superiority," is a recurrent theme. De Moivre suggests that "we may imagine Chance and Design to be, as it were, in Competition with each other" (De Moivre 1756, p. v). "Chance" and "Design" here no longer mean the presence and absence of a stochastic element, but a lack of uniformity in the probability distribution. Answering Nicholas Bernoulli, De Moivre says yes, Arbuthnot's birth data are consistent with an 18:17 ratio, but "this Ratio once discovered, and *manifestly serving to a wise purpose*, we conclude the Ratio itself, or if you will the *Form of the Die*, to be an Effect of *Intelligence and Design*" (De Moivre 1756, p. 253).

Uniformity in distribution was to be increasingly equated with absence of design or law, departure from uniformity with their presence. A famous example is Michell's argument in 1767 that optically double or multiple stars were physically so. Michell calculated that the observed clustering of stars in the heavens exceeded what could reasonably be expected if the stars were distributed at random (i.e., uniformly) over the celestial sphere, inferring "either design, or some general law" due to "the greatness of the odds against things having been in the present situation, if it was not owing to some such cause" (Michell 1767, p. 243). Michell's argument was the focus of debate for a brief period during the middle of the 19th century, a key issue being precisely this equation of uniformity with absence of law.[15]

The elements of a fundamental probability set enjoy this status for reasons which are both *aleatory* (i.e., physical or objective) and *epistemic*. The dichotomy between chance and design involves primarily the aleatory aspect of the FPS. Throughout the 18th century, the elements of an FPS were often defined in terms of equipossibility, a terminology which, as Hacking notes (1975, Chapter 14), permitted a blurring of the aleatory and epistemic aspects. The literature of the period furnishes many instances of this duality. In the *Ars Conjectandi*, for example, James Bernoulli refers to cases which are "equally possible, that is to say, each can come about as easily as any other" (*omnes casus aeque possibiles esse, seu pari facilitate evenire posse*). Laplace, on the other hand, in his *Essai philosophique*, states the following famous – and purely epistemic – criterion:

The theory of chance consists in reducing all the events of the same kind to a certain number of cases equally possible, that is to say, to such as we may be equally undecided about in regard to their existence. . . . [Laplace 1952, p. 6]

If [the various cases] are not [equally possible], we will determine first their respective possibilities, whose exact appreciation is one of the most delicate points of the theory of chance. [Laplace 1952, p. 11]

To assign equal probability to cases "such as we may be equally undecided about" is the notorious *principle of insufficient reason*. Although Laplace did not view it as controversial, many in the nineteenth century did. What determines when cases are equally probable, possible, or likely? This epistemological ambiguity in the meaning and determination of an FPS led inevitably to controversy in its application.

2. *D'Alembert and De Morgan.* For example, what is the chance of getting at least one head in two tosses of a fair coin? The standard solution to this problem regards the four possible outcomes of tossing a coin twice – HH, HT, TH, TT – as equally likely; since three out of these four cases are favorable, the probability is 3/4. In 1754, however, the French *philosophe* Jean Le Rond D'Alembert (1717–1783) advanced a different solution in his article 'Croix ou pile' in the *Encyclopédie*. D'Alembert reasoned that one would stop tossing the coin as soon as the desired head came up, so that there are really only three possible outcomes – H, TH, TT – two of which are favorable, and hence the probability is 2/3.

D'Alembert was far from being the first distinguished mathematician to make an elementary error of this type, but he is perhaps unique in the doggedness with which he subsequently defended his answer. Indeed, this was only the first of several instances where D'Alembert was led to disagree with the standard answers of the calculus of probabilities, and "with this article, the renowned mathematician opened a distinguished career of confusion over the theory of probabilities" (Baker 1975, p. 172).[16]

D'Alembert's criticisms were largely greeted with scorn and ridicule, but seldom seriously discussed. Laplace, for example, remarks that the probability would indeed be 2/3 "if we should consider with D'Alembert these three cases as equally possible . . . " (1952, p. 12), but he limits himself to giving the standard calculation without explaining why one set of equipossible cases is preferable to another.

The D'Alembert fallacy is possible because of the ambiguity in the concept of equipossibility and the Laplacean definition of probability. Laplace's treatment of these questions, although not confused, fails to come to grips with the fundamental issues. For one of the few serious discussions of D'Alembert's argument, one must turn to the writings of Augustus De Morgan, Laplace's most enthusiastic and influential English expositor during the first half of the nineteenth century.

De Morgan argued that there are essentially two very distinct considerations involved in the assessment of numerical probabilities. The first of these is *psychological*: the measurement and comparison of "the impressions made on our minds by different prospects," as in a judgment of equiprobability among alternatives. The second is *mathematical*: the rational use of such measures or comparisons, as in the computation of the probability of a complex event involving simpler, equiprobable outcomes. The two questions differ in that "any given answer to the first may admit of dispute," while "there is no fear of mathematics failing us in the second," (De Morgan 1845, p. 395).

Armed with this distinction, De Morgan was able to analyze the D'Alembert fallacy:

[W]ith regard to the objection of D'Alembert . . . , we must observe that if any individual really feel himself certain, in spite of authority and principle, as here laid down, that the preceding cases are equally probable, he is fully justified in adopting 2/3 instead of 3/4, till he see reason to the contrary, which it is hundreds to one he would find, if he continued playing for a stake throughout a whole morning, that is, accepting bets of two to one that H would not come up once in two throws, instead of requiring three to one. . . . The individual just supposed, has applied correct mathematics to a manner in which he feels obliged to view the subject, in which we think him wrong, but the error is in the first of the two considerations [above], and not in the second. [De Morgan 1845, p. 401]

Despite its interest, De Morgan's discussion is ultimately unsatisfactory. The choice of an FPS is described as a psychological consideration (which would suggest a subjectivist viewpoint), but the phrase "in which we think him wrong" suggests an objectivistic one. De Morgan appeals to experience to justify the classical choice of FPS in the D'Alembert problem, although probabilities for De Morgan were degrees of belief rather than empirical frequencies. The Laplacean view of probability was one of rational degree-of-belief, but his followers were understandably reluctant to uncouple probability from frequencies although, not surprisingly, unable to provide a logical description of the choice of FPS.

De Morgan later returned to the D'Alembert example in his *Formal Logic* (1847, pp. 199–200), and his brief discussion there is also interesting:

[I]t may happen that the state of mind which *is*, is not the state of mind which should be. D'Alembert believed that it was *two* to *one* that the first head which the throw of a halfpenny was to give would occur before the third throw; a juster view of the mode of applying the theory would have taught him it was *three* to *one*. But he *believed* it, and thought he could show reason for his belief: to him the probability *was* two to one. But I shall say, for all that, that the probability *is* three to one: meaning, that in the universal opinion of those who examine the subject, the state of mind to which a person *ought* to be able to bring himself is to look three times as confidently upon the arrival as upon the non-arrival.

When De Morgan says that, for D'Alembert, "the probability *was*," the word probability is being used in a psychological or personalist sense; when he says "the probability *is*," the sense is logical or credibilist. But to say that the probability is three to one because that is "the universal opinion of those who examine the subject," while certainly candid, is hardly a devastating refutation of D'Alembert.

De Morgan deserves considerable credit for distinguishing between the psychological process of identifying a set of outcomes as equipossible, and the mathematical use of such a set to calculate probabilities, as well as his (implicit) distinction between the subjective and objective senses of probability. Where he fails is in his account of why the probability "is" three to one, and what empirical justification, if any, such a statement requires. These, however, were basic questions for which the theory of his day had no answer.

In the latter half of the nineteenth century, a serious attack was mounted on epistemic probability and the principle of insufficient reason, and a direct confrontation with such questions could no longer be avoided.

6. THE PRINCIPLE OF INSUFFICIENT REASON

The contributions of Laplace represent a turning point in the history of probability. Before his work, the mathematical theory was (with the exception of the limit theorems of Bernoulli and DeMoivre) relatively unsophisticated, in effect a subbranch of combinatorics; its serious applications largely confined to games of chance and annuities. All this changed with Laplace. Not only did he vastly enrich the mathematical theory of the subject, both in the depth of its results and the range of the technical tools it employed, he demonstrated it to be a powerful instrument having a wide variety of applications in the physical and social sciences. Central to his system, however, was the use of the so-called principle of insufficient reason.[17]

The nineteenth century debate about the validity of the principle of insufficient reason involved, of necessity, much broader issues. Is probability empirical or epistemic in nature? Can a probability be meaningfully assigned to any event? Are all probabilities numerically quantifiable? Beginning in the 1840s, and continuing on into the twentieth century, a number of eminent British mathematicians, philosophers, and scientists began to address such questions, including De Morgan, Ellis, Mill, Forbes, Donkin, Boole, Venn, Jevons, MacColl, Edgeworth, Keynes, Ramsey, Jeffreys, and Broad.

1. *Donkin*. A comprehensive discussion of this literature would be beyond the scope of the present paper. Instead, we will confine our attention primarily to the contributions of William Fishburn Donkin, Savilian Professor of Astronomy at the University of Oxford from 1842 to 1869. Donkin wrote two papers on mathematical probability. One of these concerned the justification for the method of least squares and, although a valuable contribution to that subject, will not concern us here. The other paper is modestly titled, 'On Certain Questions Relating to the Theory of Probabilities' (Donkin 1851). Donkin's paper, although little known, is a lucid and careful attempt to clarify the foundations of the subject. It was written in response to criticisms by Forbes and others of Michell's argument that stars that are optically double are also physically so.

Donkin begins by stating that

It will, I suppose, be generally admitted, and has often been more or less explicitly stated, that the subject matter of calculation in the mathematical theory of probabilities is *quantity of belief*.

There were some dissenters to this view of probability at the time Donkin wrote (e.g., Ellis 1844; Mill 1843), but they were few in number and, due at least in part to the influence of De Morgan, Laplace's views held sway in England.[18]

Donkin's philosophical view of probability may be summarized as *relative, logical, numerical, and consistent*. Probability is relative in the sense that it is never "inherent in the hypothesis to which it refers," but "always *relative* to a state of knowledge or ignorance." Nevertheless, Donkin was not a subjectivist, because he also believed probability to be

absolute in the sense of not being relative to any individual mind; since, the same information being presupposed, all minds *ought* to distribute their belief in the same way.

Ultimately, any such theory of logical probability must resort to the principle of insufficient reason, and Donkin's was no exception. Indeed, if anything he saw its role as even more central to the theory than did Laplace:

. . . the law which must always be made the foundation of the whole theory is the following: – *When several hypotheses are presented to our mind, which we believe to be mutually exclusive and exhaustive, but about which we know nothing further, we distribute our belief equally amongst them.*

Although Boole's detailed and influential criticism of the appeal to insufficient reason was still several years off (Boole 1854, pp. 363–375), Robert Leslie Ellis had already attacked its use on the grounds that it "erected belief upon ignorance" (Ellis 1850, p. 325). Donkin's response was to stake out a limited claim for the theory:

[The force of] the argument commonly called the "sufficient reason"...in all cases depends (as it seems to me) upon a previous *assumption* that *an intelligible law exists* concerning the matter in question. If this assumption be admitted, and if it can be shown that there is only *one* intelligible law, then that must be the actual law.... A person who should dispute the propriety of dividing our belief equally amongst hypotheses about which we are equally ignorant, ought to be refuted by asking him to state *which is to be preferred*. He must either admit the proposed law, or maintain that there is no law at all.

This observation would not have disarmed Ellis, Boole, or Venn, who indeed denied the existence of any (determinate in the case of Boole) law at all. But it did draw the line clearly. Its vulnerability, as Boole realized, is simply that two or more sets of "mutually exclusive and exhaustive" hypotheses may present themselves "about which we know nothing further," and which give rise to incompatible probability assignments. Ramsey saw it as a virtue of the subjectivistic theory that it eluded this dilemma by dispensing with the requirement of a unique law, admitting more than one probability assignment as possible (Ramsey 1926, pp. 189–190).

But can one calculate probabilities no matter how complex the setting or information available? Cournot, for example, had earlier argued that there were three distinct categories of probability – objective, subjective, and philosophical, the last involving situations whose complexity precluded mathematical measurement.[19]

Donkin thought such arguments, essentially pragmatic in nature, not to the point:

...I do not see on what ground it can be doubted that every definite state of belief concerning a proposed hypothesis is in itself capable of being represented by a numerical expression, however difficult or impracticable it may be to ascertain its actual value.... [It is important to distinguish] the difficulty of *ascertaining numbers* in certain cases from a supposed difficulty of *expression by means of numbers*. The former difficulty is real, but merely relative to our knowledge and skill; the latter, if real, would be absolute, and inherent in the subject matter, which I conceive not to be the case.

This was an important distinction. It expresses a tenet of faith of logical probability: that all probabilities can, in principle, be measured. On a basic philosophical level, such theories have never really answered Ramsey's simple criticism:

It is true that about some particular cases there is agreement, but these somehow paradoxically are always immensely complicated; we all agree that the probability of a coin coming down heads is 1/2, but we can none of us say exactly what is the evidence which forms the other term for the probability relation about which we are then judging. If, on the other hand, we take the simplest possible pairs of propositions such as 'This is red', and 'That is blue', or 'This is red' and 'That is red', whose logical relations should surely be easiest to see, no one, I think, pretends to be sure what is the probability relation between them. [Ramsey 1926]

2. *Boole*. The first influential critic of the principle of insufficient reason was Boole. He says of its derivation:

It has been said, that the principle involved in the above and in similar applications is that of the equal distribution of our knowledge, or rather of our ignorance – the assigning to different states of things of which we know nothing, and upon the very ground that we know nothing, equal degrees of probability. I apprehend, however, that this is an arbitrary method of procedure. [Boole 1854, p. 370]

As we have seen earlier (Section 3), to justify his criticism Boole pointed to instances where it was possible to partition the sample space of possible outcomes in different ways, each of which could plausibly be viewed as equipossible. Boole's criticisms, unfortunately, became more confusing as he attempted to clarify them. One might be forgiven, for example, for interpreting the passage just quoted as a clear rejection of the principle. But Boole later wrote:

I take this opportunity of explaining a passage in the *Laws of Thought*, p. 370, relating to certain applications of the principle. Valid objection lies not against the principle itself, but against its application through arbitrary hypotheses, coupled with the assumption that any result thus obtained is necessarily the true one. The application of the principle employed in the text and founded upon the general theorem of development in Logic, I hold to be *not* arbitrary. [Boole 1862]

Perusal of "the application of the principle employed in the text" reveals it to be of the balls in an urn type, and what Boole now appears to be defending might be called the principle of *cogent* reason: if one possesses some information about the different alternatives, but this information is equally distributed amongst them, then one is justified in assigning the alternatives equal probability.

Boole appeared to regard both probabilistic independence (which he used extensively in his system) and uniformity of distribution as assumptions of *neutrality*, in each case a *via media* between conflicting extremes. There is a simple geometric sense in which this is true for the assumption of uniformity: the uniform distribution on $n + 1$ elements is the barycenter of the n-dimensional simplex of all probability distributions. But once more the consequences of a symmetry assumption lurk only partially visible. For depending on the use being made of a probability distribution, symmetrical or uniform distributions can often represent an extreme type of behavior. A good example of this involves the "birthday paradox": in a group of 23 or more people, the odds exceed 1/2 that at least two persons share a birthday in common (Feller 1968, p. 33). The calculation on which this statement is based assumes that births occur uniformly throughout the year. Although empirically false (e.g., Izenman and Zabell 1981), this does not affect the validity of the conclusion: the probability of a birthday "match" is minimized when the distribution of births is uniform (so that the probability of a match will be even greater under the true distribution).

It is difficult to assess Boole's immediate impact on his contemporaries. As the distinguished author of *The Laws of Thought*, his views on probability were certainly treated with respect. Nevertheless, they were highly idiosyncratic and confused in important respects.[20] Given the complexity and unattractiveness of his own system, and lacking the alternative philosophical foundation to the Laplacean edifice that was later provided by Venn's *Logic of Chance*, there was an obvious reluctance to abandon the classical theory. Nevertheless, his pointing to the fundamental ambiguity in the principle of insufficient reason was a lasting contribution, remembered long after the rest of his work on probability was forgotten.

Donkin represents what may be the highwater mark in the defense of the Laplacean position; Boole was its first influential English critic. After Boole and Venn the Laplaceans were on the defensive, first in the philosophical, later in the statistical and scientific communities. In response to the criticisms of Boole and his successors, many attempts were made to state unambiguous formulations of the principle of insufficient reason (e.g., by von Kries and Keynes), but their increasing obscurity and complexity ensured their rejection.[21]

The debate about the principle of insufficient reason and its consequence, Laplace's rule of succession, tapered off in the 1920s. This was partly because Ramsey's 1926 essay 'Truth and Probability' made the principle superfluous as a foundation for epistemic probability. When Fisher and Neyman produced statistical methodologies independent of the Bayes–Laplace edifice, Bayesian statistics essentially disappeared, only to be resuscitated by Savage nearly a quarter of a century later with the publication in 1954 of his *Foundations of Statistics.*

Savage's conversion to subjectivism occurred after he became acquainted with de Finetti's work, and his writings were largely responsible for bringing it into the mainstream of philosophical and statistical thought. At the center of de Finetti's system was the notion of exchangeability, and thus, initially exorcised, symmetry re-entered epistemic probability.

7. WHAT IS TO BE DONE?

Symmetry arguments are tools of great power; therein lies not only their utility and attraction, but also their potential treachery. When they are invoked one may find, as did the sorcerer's apprentice, that the results somewhat exceed one's expectations. Nevertheless, symmetry arguments enjoy an honored and permanent place in the arsenal of probability. They underlie the classical definition of probability that held sway for over two centuries, are central to virtually all quantitative theories of induction, appear as exchangeability assumptions in subjectivist theories, and, in the guise of group-invariance, still play an important role in modern theoretical statistics. Their use calls for judicious caution rather than benign neglect.

The ambiguity underlying the proper role of symmetry assumptions in the theory of probability stems in part from a corresponding ambiguity about the role the axioms play in the various axiomatic formulations of probability. Do the axioms enjoy a privileged status *vis-à-vis* their deducible consequences? Are they supposed to be intuitively more evident or simpler in form? If the justification for the axioms is their intuitive acceptability, what if some of their consequences violate those intuitions? As in so many cases, one can identify two polar positions on such issues, that of the *left-wing dadaists* and the *right-wing totalitarians*.[22]

The left-wing dadaists not only demand that the axioms be grounded in our intuitions, but that *all* deducible consequences of the axioms must be intuitively acceptable as well. Intuitive acceptability was the warrant for the axioms in the first place, and since there is no obvious reason to favor certain intuitions over others, all must be satisfied. If the consequences of a set of axioms violate our intuitions, then those axioms must be abandoned and replaced. A leading exponent of this position is L. Jonathan Cohen.[23]

27

The problem with such a position is that our intuitions, or at least our *untutored* intuitions, are often mutually inconsistent and any consistent theory will necessarily have to contradict some of them. During the last two decades many psychologists, notably Daniel Kahneman and Amos Tversky, have demonstrated that popular intuitions are often inconsistent not merely with the standard axioms of probability, but with essentially *any* possible axiomatization of probability; that "people systematically violate principles of rational decision-making when judging probabilities, making predictions, or otherwise attempting to cope with probabilistic tasks" (Slovic, Fischhoff, and Lichtenstein 1976).[24]

The right-wing totalitarians, on the other hand, believe that once an axiom system is adopted, one must accept without question every consequence that flows from it. One searches within one's heart, discovers the basic properties of belief and inference, christens them axioms, and then all else follows as logical consequence. Once the axioms are brought to the attention of unbelievers, they must, like Saul on the road to Damascus, be smitten by instantaneous conversion or they stand convicted of irrational obtuseness. One representative of this position is E. T. Jaynes, who dates his adherence to Bayesianism to the time when he encountered Cox's axiomatization of epistemic probability, and who views the Shannon axioms for entropy as an unanswerable foundation for his method of maximum entropy.[25]

This position errs in giving the axioms too distinguished a position, just as the previous position gave them too little. A set of axioms A, together with $T(A)$, the theorems deducible from it, forms a self-consistent whole S. Let us say that any subset $B \subseteq S$, such that $B \cup T(B) = S$, is an *axiom-system for S*. Mathematically speaking, all possible axiom-systems for S must be regarded as starting out on an equal footing, and which axiom-system is ultimately chosen is essentially a matter of preference, depending on considerations such as simplicity, elegance, and intuitive acceptability.

The key point is that having tentatively adopted an axiom system, one is not obligated to uncritically accept its consequences. In both formal logic and subjective probability, the theory polices sets of beliefs by testing them for inconsistencies, but it does not dictate how detected inconsistencies should be removed. If, as was the case with some of the symmetry assumptions previously discussed, the consequences are deemed unacceptable, then the assumption will be discarded. If, on the other hand, the axioms seem compelling, as in mathematical probability, then surprising consequences such as the birthday paradox will be regarded as valuable correctives to our erroneous, untutored intuitions; that is why the theory is useful. What is or should be at play is a dynamic balance. As Nelson Goodman argues:

> Inferences are justified by their conformity to valid general rules, and . . . general rules are justified by their conformity to valid inferences. But this circle is a virtuous one. The point is that rules and particular inferences alike are justified by being brought into agreement with each other. *A rule is amended if it yields an inference we are unwilling to accept; an inference is rejected if it violates a rule we are unwilling to amend.* The process of justification is the delicate one of making mutual adjustments between rules and accepted inferences; and in the agreement achieved lies the only justification needed for either [Goodman 1979, p. 64].

Symmetry assumptions must therefore be tested in terms of the particular inferences they give rise to. But – and this is the rub – particular inferences can only be reasonably judged in terms of particular situations, whereas symmetry assumptions are often proposed in abstract and theoretical settings devoid of concrete specifics.[26]

Fundamentally at issue here are two very different approaches to the formulation of a logic of probability. Extreme subjectivists adopt a *laissez faire* approach to probability assignments, emphasizing the unique aspects attending the case at hand. They do not deny the utility of symmetry arguments, but, as Savage remarks, they "typically do not find the contexts in which such agreement obtains sufficiently definable to admit of expression in a postulate" (Savage 1954, p. 66). Such arguments fall instead under the rubric of what I. J. Good terms "suggestions for using the theory, these suggestions belonging to the technique rather than the theory" itself (Good 1952, p. 107).

Proponents of logical theories, in contrast, believe (at least in principle) that if the evidence at one's disposal is stated with sufficient precision in a sufficiently rich language then agreement can be forced via considerations of symmetry. At the level of ordinary language such claims founder at the very outset on Ramsey's simple objection (quoted earlier in Section 6). Instead, simple model languages are introduced and probabilities computed "given" statements descriptive of our state of knowledge. Such formal systems do not escape subjectivism, they enshrine it in the equiprobable partitions assumed.

Practical attempts to apply logical probability always seem to lead back to discussions about events "concerning the probability of which we absolutely know nothing antecedently to any trials made concerning it." Such attempts are ultimately divorced from reality, if only because understanding the very meaning of the words employed in describing an event already implies knowledge about it. Thus, it is not surprising that the three leading twentieth century proponents of logical probability – Keynes, Jeffreys, and Carnap – all eventually recanted to some extent or another.[27] Carnap, for example, wrote

I think there need not be a controversy between the objectivist point of view and the subjectivist or personalist point of view. Both have a legitimate place in the context of our work, that is, the construction of a system of rules for determining probability values with respect to possible evidence. At each step in the construction, a choice is to be made; the choice is not completely free but is restricted by certain boundaries. Basically, there is merely a difference in attitude or emphasis between the subjectivist tendency to emphasize the existing freedom of choice, and the objectivist tendency to stress the existence of limitations. [Carnap 1980, p. 119]

This little-known, posthumously published passage is a substantial retreat from the hard-core credibilism of the *Logical Foundations of Probability*. But it was inevitable. Symmetry arguments lie at the heart of probability. But they are tools, not axioms, always to be applied with care to specific instances rather than general propositions.

8. ENVOI

As a final illustration of the seductive nature of symmetry arguments in probability, and as a challenge to the reader, I end with a little puzzle, which I will call the *exchange paradox*:[28]

A, B, and C play the following game. C acts as referee and places an unspecified amount of money x in one envelope and amount $2x$ in another envelope. One of the two envelopes is then handed to A, the other to B.

A opens his envelope and sees that there is $10 in it. He then reasons as follows: "There is a 50–50 chance that B's envelope contains the lesser amount x (which would therefore be $5), and a 50–50 chance that B's envelope contains the greater amount $2x$ (which would therefore be $20). If I exchange envelopes, my expected holdings will be $(1/2)\$5 + (1/2)\$20 = \$12.50$, $2.50 in excess of my present holdings. Therefore I should try to exchange envelopes."

When A offers to exchange envelopes, B readily agrees, since B has already reasoned in similar fashion.

It seems unreasonable that the exchange be favorable to both, yet it appears hard to fault the logic of either. I will resist the temptation to explain what I take to be the resolution of the paradox, other than noting that all hinges on A's apparently harmless symmetry assumption that it is equally likely that B holds the envelope with the greater or the lesser amount.[29]

NOTES

1. The symbol $\binom{n}{k}$ denotes the binomial coefficient $n!/[k!(n-k)!]$. Note that in the theorem the sequence is assumed to be infinite; this requirement is sometimes overlooked, although it is necessary for the general validity of the theorem.
2. There also exist finite forms of de Finetti's theorem, which permit one to dispense with the assumption that the number of trials is infinite. In such cases the integral mixture is either replaced by a discrete sum or serves as an approximation to the exact probability; see Diaconis and Freedman (1980a).
3. The existence of limiting frequencies for infinite exchangeable sequences follows from their stationarity, and is an immediate consequence of the ergodic theorem; see, e.g., Breiman (1968, p. 118, Theorem 6.28).
4. For further discussion of Bayes's scholium, see Murray (1930), Edwards (1978). For an interesting account of how Bayes's argument has often been misconstrued by statisticians to fit their foundational preconceptions, see Stigler (1982).
5. It is an interesting historical footnote that Johnson's derivation almost never appeared. After the appearance of the third volume of his *Logic* in 1924, Johnson began work on a fourth volume, to be devoted to probability. Unfortunately, Johnson suffered a stroke in 1927, and the projected work was never finished. Drafts of the first three chapters were edited by R. B. Braithwaite and published posthumously as three separate papers in *Mind* during 1932. Johnson's mathematical derivation of the continuum of inductive methods from the sufficientness postulate appeared as an appendix in the last of the three. G. E. Moore, then editor of *Mind*, questioned whether so technical a result would be of general interest to its readership, and it was only on the insistence of Braithwaite that the appendix was published (Braithwaite 1982, personal communication).
6. For further information about Johnson's sufficientness postulate, and a complete version of his proof, see Zabell (1982).
7. In brief, this is because even when one lacks information about specific, identifiable categories, one may possess information about the vector of *ordered* probabilities. (For example, one may know that a die is biased in favor of one face, but not know which face it is.)
8. See generally Heath (1913, Chapter 4); Kahn (1960); Dicks (1970, Chapter 3). For the original Greek texts of the fragments of Anaximander, with accompanying English translation, commentary, and discussion, see Kirk and Raven (1957, Chapter 3).
9. Perhaps the most pessimistic assessment of the state of our information is that of Dicks (1970, pp. 45–46).
10. In its general form (neither of two exactly symmetrical alternatives will occur), it also crops up from time to time in 19th century philosophical discussions of probability. Two examples are (1) Bolzano: "... if we are to have a rational expectation that a certain result will take place, for example that Caius will draw a certain ball from several balls in an urn, then we must presuppose that the relation between these balls and Caius is such that the reasons for drawing that particular ball are not exactly like the reasons for drawing some other ball, since otherwise he wouldn't draw any" (Bolzano 1837, p. 245 of 1972 edition.); (2) Cook Wilson: "... if a number of cases, mutually exclusive ..., were in the nature of things equally possible, not

one of them could happen. If the claim of any one of them in reality were satisfied, so must the claim of any other, since these claims are equal, and therefore if one happens all must, but by hypothesis if one happens no other can; thus the only possible alternative is that none of them can happen" (Wilson 1900, p. 155).

11. Aristocles, quoted in Long (1974, p. 81); c.f. Diogenes Laertius, *Life of Pyrrho* 9.107; Sextus Empiricus, *Outlines of Pyrrhonism* 1.8. For general information on the Pyrrhonian skeptics, see Stough (1969, Chapter 2); Long (1974, pp. 75–88). The *ou mallon* argument itself is discussed in some detail by DeLacy (1958).

12. See generally Stough (1969, pp. 50–66); Long (1974, pp. 95–99).

13. See, e.g., Plato, *Laws* 709, 889 b–d; Aristotle, *Metaphysics* 1070ab. (Strictly speaking, Aristotle distinguishes between *automaton* (chance, spontaneity) and *tyche* (luck, fortune).

14. For further discussion of Arbuthnot, see Hacking (1965, pp. 75–77); Hacking (1975, Chapter 18); Pearson (1978, pp. 127–133, 161–162).

15. For a recent and very readable account of the dispute, see Gower (1982). Similar issues arose in later discussions of geometrical probability: what does it mean to select points (or lines, or triangles) at random? Venn (1888, pp. 100–101), reporting one such discussion, quotes the English mathematician Crofton as asserting that "at random" has "a very clear and definite meaning; one which cannot be better conveyed than by Mr Wilson's definition, 'according to no law'. . . . " "Mr. Crofton holds," Venn continues, "that any kind of *unequal* distribution [of points in a plane] would imply law," to which Venn retorts, "Surely if they tend to become *equally* dense this is just as much a case of regularity or law." Where James Bernoulli had attempted to subsume the probability of causes under that of chances (to use Hume's terminology), the frequentist Venn subsumes the probability of chances under that of causes.

16. See generally Todhunter (1865, Chapter 13); Baker (1975, pp. 171–180); Pearson (1978, Chapter 12). For a recent but unconvincing attempt at rehabilitation, see Daston (1979).

17. Laplace nowhere actually uses this term, which is of later origin. Writing in 1862, Boole refers to "that principle, more easily conceived than explained, which has been differently expressed as the 'principle of non-sufficient reason', the principle of equal distribution of knowledge or ignorance' [footnote omitted], and the 'principle of order'," (Boole 1862).

18. When Donkin wrote his paper the first frequentist theories (apart from occasional allusions in the earlier literature) were less than a decade old. As Porter (1986, p. 77) notes, "in 1842 and 1843, four writers from three countries independently proposed interpretations of probability that were fundamentally frequentist in character." These four – Jakob Friedrick Fries in Germany, Antoine Augustin Cournot in France, and Richard Leslie Ellis and John Stuart Mill in England – were the harbingers of an increasingly empirical approach to probability. (Curiously, after correspondence with the astronomer John Herschel, Mill actually withdrew his objections to Laplace's epistemic view of probability from the second (1846) and later editions of his *Logic*; see Strong (1978).) Despite this early efflorescence, the frequency theory did not begin to gain widespread acceptance until its careful elaboration, nearly a quarter of a century later, in John Venn's *Logic of Chance* (1st ed. 1866). For discussion of the work of Fries, Cournot, Ellis, and Mill, see Porter

(1986, pp. 77–88), Stigler (1986, pp. 195–200); for discussion of Venn's *Logic*, Salmon (1980).

19. The argument that some probabilities are "philosophical" (i.e., inherently non-numerical) was often made by those who thought the mathematical theory had outreached its grasp. Strong (1976, p. 207, n. 5) notes the use of the distinction in K. H. Frömmichen's 1773 work, *Ober die Lehre des Wahrscheinlich*, "the earliest . . . that I have been able definitely to date," as well as the better known treatment in Kant's *Logik* of 1781. See von Wright (1957, p. 217, n. 9) for further references to the 19th century literature. In addition to the names given there, one could add those of the Scottish philosopher Dugald Stewart and the English jurists Starkie, Wills, and Best. For the related criticisms of the French positivists Destutt de Tracy, Poinsot, and Comte, see Porter (1986, p. 155) and Stigler (1986, pp. 194–195).

20. Many of these are touched on by Keynes in scattered passages throughout his *Treatise on Probability* (1921). Hailperin (1976) is a useful attempt at rational reconstruction. For discussion of Boole's criticism of the Laplace/De Morgan analysis of inductive reasoning in terms of probability, see the excellent article of Strong (1976).

21. See generally Keynes (1921, Chapters 4 and 6).

22. There is obviously an element of intentional caricature in what follows, although perhaps less than might be supposed.

23. " . . . ordinary human reasoning . . . cannot be held to be faultily programmed: it sets its own standards" (Cohen 1981, p. 317).

24. Much of this work is summarized in Kahneman, Slovic, and Tversky (1982).

25. Although not readily available, Jaynes's early Socony Mobil Oil lecture notes (Jaynes 1958) provide a vigorous and very readable exposition of his viewpoint.

26. There are some notable exceptions to this. W. E. Johnson, for example, in discussing his sufficientness postulate, argued that:

"the postulate adopted in a controversial kind of theorem cannot be generalized to cover all sorts of working problems; so it is the logician's business, having once formulated a specific postulate, to indicate very carefully the factual and epistemic conditions under which it has practical value." (Johnson 1932, pp. 418–419)

27. For Keynes's recantation, see Good (1965, p. 7). In the third edition of his book *Scientific Inference*, Jeffreys suggests that in controversial cases the appropriate choice of reference prior could be decided by an international panel of experts. Such a position is obviously incompatible with credibilism as usually understood. For Carnap, see the text *infra*.

28. I first heard the paradox from Steve Budrys of the Odesta Corporation, on an otherwise unmemorable night at the now defunct Chessmates in Evanston. It does not originate with him, but I have been unable to trace its ultimate source.

    Note added in proof: Persi Diaconis and Martin Gardner inform me that the paradox is apparently due to the French mathematician Maurice Kraitchik; see Maurice Kraitchik, *Mathematical Recreations*, 2nd ed. (New York: Dover, 1953), pp. 133–134. In Kraitchik's version two persons compare their neckties, the person with the less valuable necktie to receive both.

29. I thank Persi Diaconis, David Malament, and Brian Skyrms for helpful comments.

33

REFERENCES

Arbuthnot, John (1711). 'An argument for divine providence taken from the constant regularity observed in the births of both sexes', *Philosophical Transactions of the Royal Society of London* **27**, 186–190.

Baker, Keith Michael (1975). *Condorcet: From Natural Philosophy to Social Mathematics* (Chicago: University of Chicago Press).

Bayes, Thomas (1764). 'An essay towards solving a problem in the doctrine of chances', *Philosophical Transactions of the Royal Society of London* **53**, 370–418.

Bolzano, Bernard (1837). *Wissenschaftslehre*. Translated 1972 under the title *Theory of Science* (R. George, ed. and trans.) (Berkeley and Los Angeles: University of California Press).

Boole, George (1854). *An Investigation of the Laws of Thought* (London: Macmillan.) (Reprinted 1958, New York: Dover Publications.)

Boole, George (1862). 'On the theory of probabilities', *Philosophical Transactions of the Royal Society of London* **152**, 386–424.

Breiman, Leo (1968). *Probability* (Reading, Mass.: Addison-Wesley).

Broad, C. D. (1975). *Leibniz: An Introduction* (Cambridge, UK: Cambridge University Press).

Carnap, Rudolph (1950). *Logical Foundations of Probability* (Chicago: The University of Chicago Press. Second edition, 1960).

Carnap, Rudolph (1952). *The Continuum of Inductive Methods* (Chicago: University of Chicago Press).

Carnap, Rudolph (1980). 'A basic system of inductive logic, part II', in *Studies in Inductive Logic and Probability*, volume II (Richard C. Jeffrey, ed.) (Berkeley and Los Angeles: University of California Press) pp. 7–155.

Cohen, L. Jonathan (1981). 'Can human irrationality be experimentally demonstrated?', *The Behavioral and Brain Sciences* **4**, 317–370 (with discussion).

Cournot, Antoine Augustin (1843). *Exposition de la théorie des chances et des probabilites* (Paris: Libraire de L. Hachette).

Daston, Lorraine J. (1979). 'D'Alembert's critique of probability theory', *Historia Mathematica* **6**, 259–279.

De Finetti, Bruno (1937). 'La prévision: ses lois logiques, ses sources subjectives', *Annales de l'Institut Henri Poincaré* **7**, 1–68.

DeLacy, Phillip (1958). 'Ou mallon and the antecedents of ancient scepticism', *Phronesis* **3**, 59–71.

De Moivre, Abraham (1756). *The Doctrine of Chances* (3rd ed.), London.

De Morgan, Augustus (1845). 'Theory of probabilities', *Encyclopedia Metropolitana, Vol. 2: Pure Mathematics* (London: B. Fellowes *et al.*) pp. 393–490.

De Morgan, Augustus (1847). *Formal Logic: Or the Calculus of Inference Necessary and Probable*. London.

Diaconis, Persi (1977). 'Finite forms of de Finetti's theorem on exchangeability', *Synthese* **36**, 271–281.

Diaconis, Persi and Freedman, David (1980a). 'Finite exchangeable sequences', *Annals of Probability* **8**, 745–764.

Diaconis, Persi and Freedman, David (1980b). 'De Finetti's theorem for Markov chains', *Annals of Probability* **8**, 115–130.

Diaconis, Persi and Freedman, David (1980c). 'De Finetti's generalizations of exchange-ability', in *Studies in Inductive Logic and Probability*, volume II (Richard C. Jeffrey, ed.) (Berkeley and Los Angeles: University of California Press) pp. 233–249.

Diaconis, Persi and Freedman, David (1985). 'Partial exchangeability and sufficiency', *Statistics: Applications and New Directions*. In *Proceedings of the Indian Statistical Institute Golden Jubilee International Conference* (Calcutta: Indian Statistical Institute) pp. 205–236.

Dicks, D. R. (1970). *Early Greek Astronomy to Aristotle* (Ithaca: Cornell University Press).

Donkin, William Fishburn (1851). 'On certain questions relating to the theory of probabilities', *Philosophical Magazine* **1**, 353–368; **2**, 55–60.

Edwards, A. W. F. (1978). 'Commentary on the arguments of Thomas Bayes', *Scandinavian Journal of Statistics* **5**, 116–118.

Ellis, Richard Leslie (1844). 'On the foundation of the theory of probabilities', *Transactions of the Cambridge Philosophical Society* **8**, 1–6.

Ellis, Richard Leslie (1850). 'Remarks on an alleged proof of the "method of least squares" contained in a late number of the *Edinburgh Review*', *Philosophical Magazine* **37**, 321–328.

Feller, William (1968). *An Introduction to Probability Theory and Its Applications*, vol. 1, 3rd ed. (New York: Wiley).

Good, Irving John (1952). 'Rational decisions', *Journal of the Royal Statistical Society B* **14**, 107–114.

Good, Irving John (1965). *The Estimation of Probabilities: An Essay on Modern Bayesian Methods* (Cambridge, Mass.: M.I.T. Press).

Goodman, Nelson (1979). *Fact, Fiction, and Forecast* (3rd ed.) (Indianapolis: Hackett Publishing Company).

Gower, Barry (1982). 'Astronomy and probability: Forbes versus Michell on the distribution of the stars', *Annals of Science* **39**, 145–160.

Hacking, Ian (1965). *The Logic of Statistical Inference* (Cambridge, UK: Cambridge University Press).

Hacking, Ian (1967). 'Slightly more realistic personal probability', *Philosophy of Science* **34**, 311–325.

Hacking, Ian (1975). *The Emergence of Probability* (Cambridge University Press).

Hailperin, Theodore (1976). *Boole's Logic and Probability*. Studies in Logic and the Foundations of Mathematics, volume 85 (Amsterdam: North-Holland).

Heath, Sir Thomas (1913). *Aristarchus of Samos: The Ancient Copernicus* (Oxford: The Clarendon Press). (Reprinted 1981, New York: Dover Publications.)

Hewitt, Edwin and Savage, Leonard J. (1955). 'Symmetric measures on Cartesian products', *Transactions of the American Mathematical Society* **80**, 470–501.

Hume, David (1739). *A Treatise of Human Nature*. London. (Page references are to the 2nd edition of the L. A. Selbe-Bigge text, revised by P. H. Nidditch, Oxford: The Clarendon Press, 1978.)

Hussey, Edward (1972). *The Presocratics* (New York: Charles Scribner's Sons).

Izenman, Alan J. and Zabell, Sandy L. (1981). 'Babies and the blackout: The genesis of a misconception', *Social Science Research* **10**, 282–299.

Jaynes, Edwin T. (1958). *Probability Theory in Science and Engineering*. Colloquium Lectures in Pure and Applied Science, no. 4 (Dallas: Socony Mobil Oil).

Jeffrey, Richard C. (1977). 'Mises redux', *Basic Problems in Methodology and Linguistics: Proceedings of the Fifth International Congress of Logic, Methodology and Philosophy of Science, Part III* (R. Butts and J. Hintikka, eds.) (Dordrecht: D. Reidel).

Johnson, William Ernest (1924). *Logic, Part III: The Logical Foundations of Science* (Cambridge, UK: Cambridge University Press).

Johnson, William Ernest (1932). 'Probability: The deductive and inductive problems', *Mind* **41**, 409–423.

Kahn, Charles H. (1960). *Anaximander and the Origins of Greek Cosmology* (New York: Columbia University Press).

Kahneman, D., Slovic, P., and Tversky, A. (1982). *Judgment Under Uncertainty: Heuristics and Biases* (Cambridge University Press).

Keynes, John Maynard (1921). *A Treatise on Probability* (London: Macmillan).

Kirk, G. S. and Raven, J. E. (1957). *The Presocratic Philosophers: A Critical History with a Selection of Texts* (Cambridge University Press).

Kyburg, Henry (1974). 'Propensities and probabilities', *British Journal for the Philosophy of Science* **25**, 358–375.

Laplace, Pierre Simon Marquis de (1952). *A Philosophical Essay on Probabilities* (F. W. Truscott and F. L. Emory, trans.) (New York: Dover Publications).

Long, A. A. (1974). *Hellenistic Philosophy: Stoics, Epicureans, Sceptics* (New York: Charles Scribner's Sons).

Michell, J. (1767). 'An inquiry into the probable parallax and magnitude of the fixed stars from the quantity of light which they afford to us, and the particular circumstances of their situation', *Philosophical Transactions of the Royal Society* **57**, 234–264.

Mill, John Stuart (1843). *A System of Logic*, 2 vols. London.

Murray, F. H. (1930). 'Note on a scholium of Bayes', *Bulletin of the American Mathematical Society* **36**, 129–132.

Oresme, Nicole (1966). *De proportionnibus proportionum and Ad pauca respicientes* (E. Grant ed. and trans.) (Madison: University of Wisconsin Press).

Owen, G. E. L. (1966). 'Plato and Parmenides on the timeless present', *The Monist* **50**, 317–340.

Pearson, Karl (1978). *The History of Statistics in the 17th and 18th Centuries* (E. S. Pearson, ed.) (New York: Macmillan).

Porter, Theodore (1986). *The Rise of Statistical Thinking* (Princeton: Princeton University Press).

Ramsey, Frank Plumpton (1926). 'Truth and probability', in *The Foundations of Mathematics and Other Logical Essays* (R. B. Braithwaite, ed.) (London: Routledge and Kegan Paul, 1931) pp. 156–198.

Salmon, Wesley C. (1980). 'John Venn's *Logic of Chance*', in *Pisa Conference Proceedings*, vol. 2 (J. Hintikka, D. Gruender, and E. Agazzi, eds.) (Dordrecht: D. Reidel).

Savage, Leonard J. (1954). *The Foundations of Statistics* (New York: John Wiley) (Reprinted 1972, New York: Dover).

Slovic, P., Fischhoff, B., and Lichtenstein, S. (1976). 'Cognitive processes and societal risk taking', in J. S. Carroll and J. W. Payne (eds.), *Cognition and Social Behavior* (Hillsdale, N.J.: Erlbaum).

Stigler, Stephen M. (1982). 'Thomas Bayes's Bayesian inference', *Journal of the Royal Statistical Society* Series A **145**, 250–258.

Stigler, Stephen M. (1986). *The History of Statistics* (Cambridge, MA: Harvard University Press).

Stough, Charlotte L. (1969). *Greek Skepticism: A Study in Epistemology* (Berkeley: University of California Press).

Strong, John V. (1976). 'The infinite ballot box of nature: De Morgan, Boole, and Jevons on probability and the logic of induction', *PSA 1976: Proceedings of the Philosophy of Science Association* **1**, 197–211.

Strong, John V. (1978). 'John Stuart Mill, John Herschel, and the "probability of causes" ', *PSA 1978: Proceedings of the Philosophy of Science Association*, **1**, 31–41.

Todhunter, Isaac (1865). *A History of the Mathematical Theory of Probability from the Time of Pascal to That of Laplace* (London: Macmillan) (Reprinted 1965, New York: Chelsea.)

Venn, John (1888). *The Logic of Chance* (3rd ed.) (London: Macmillan).

Wilson, John Cook (1900). 'Inverse or "a posteriori" probability', *Nature* **63**, 154–156.

von Wright, Georg Henrik (1957). *The Logical Problem of Induction* (2nd revised edition). (New York: Macmillan.)

Zabell, Sandy L. (1982). 'W. E. Johnson's "sufficientness" postulate', *Annals of Statistics* **10**, 1091–1099.

Note: Translations of Greek and Latin passages are taken unless otherwise noted from the editions in the Loeb Classical Library.

2

The Rule of Succession

1. INTRODUCTION

Laplace's rule of succession states, in brief, that if an event has occurred m times in succession, then the probability that it will occur again is $(m + 1)/(m + 2)$. The rule of succession was the classical attempt to reduce certain forms of inductive inference – "pure inductions" (De Morgan) or "eductions" (W. E. Johnson) – to purely probabilistic terms. Subjected to varying forms of ridicule by Venn, Keynes, and many others, it often served as a touchstone for much broader issues about the nature and role of probability.

This paper will trace the evolution of the rule, from its original formulation at the hands of Bayes, Price, and Laplace, to its generalizations by the English philosopher W. E. Johnson, and its perfection at the hands of Bruno de Finetti. By following the debate over the rule, the criticisms of it that were raised and the defenses of it that were mounted, it is hoped that some insight will be gained into the achievements and limitations of the probabilistic attempt to explain induction. Our aim is thus not purely – or even primarily – historical in nature.

As usually formulated, however, the rule of succession involves some element of the infinite in its statement or derivation. That element is not only unnecessary, it can obscure and mislead. We begin therefore by discussing the finite version of the rule, its statement, history, and derivation (sections 2–3), and then use it as a background against which to study the probabilistic analysis of induction from Bayes to de Finetti (sections 4–9). Sections 4–6 deal largely with historical issues; sections 7–9 matters mathematical and foundational.

2. THE FINITE RULE OF SUCCESSION

One form of enumerative induction involves performing an experiment that can, at least in principle, be repeated an indefinitely large number of times

Reprinted with permission from *Erkenntis* 31 (1989): 283–321, © 1989 by Kluwer Academic Publishers.

("trials"), with one of two possible outcomes ("success" vs. "failure"). In this case it makes sense to refer to the (unknown) probability p of success, i.e., the limiting frequency, propensity, or objective chance of success. Under the classical Laplacean analysis, if the trials are independent, and all possible values of p are assumed equally likely, then given r successes in m trials, the probability of a success on the next trial is

$$\int_0^1 p^{r+1}(1-p)^{m-r}\,dp \Big/ \int_0^1 p^r(1-p)^{m-r}\,dp = (r+1)/(m+2).$$

This is Laplace's *rule of succession*.[1]

For certain types of enumerative induction the Laplacean schema is unsatisfactory. If one is noting the color of ravens, tagging each one after its color is recorded, then the universe being sampled is finite, and the sampling is being done without replacement (i.e., each raven is observed at most once). For this reason, in 1918 the English philosopher C. D. Broad repeated the Laplacean analysis, but for sampling from a finite urn without replacement (the Laplacean picture can be thought of, in a way that can be made mathematically precise, as sampling from an urn with an infinite number of balls). Of course, there are questions about the extent to which observing the color of ravens corresponds to sampling balls from an urn (realistically, one only sees ravens in one's neighborhood) – important questions, and ones also considered by Broad – but let us set these aside for the moment and consider Broad's simple mathematical question:

Consider an urn with a finite but unknown number of balls n, each of which is either black or white. Suppose a sample of m balls is drawn at random without replacement from the urn. If nothing is known about the relative proportion of black and white balls, and all m of the balls drawn are black, what is the probability that the next ball drawn is black?

Of course, some assumption must be made about the prior probability for the proportion of blacks. The natural assumption, in analogy to the Laplacean treatment, is that all possible proportions j/n are equally likely, and this is the one that Broad made in 1918.[2] Broad discovered that, surprisingly, the answer does not depend on n, the population size, but only on m, the sample size, and that the answer is *identical* to Laplace's rule, i.e., $(m+1)/(m+2)$.

The proof is not difficult. A simple application of Bayes's theorem shows that the desired probability is

$$\frac{\sum_{j=m+1}^n j(j-1)(j-2)\ldots(j-m)}{(n-m)\sum_{j=m}^n j(j-1)(j-2)\ldots(j-m+1)}.$$

The problem thus reduces to the evaluation of two sums, and, as Broad notes, "it can easily be shown that" their ratio is $(m + 1)(m + 2)$. If the sum in the denominator is denoted $S_{m,n}$, then a simple inductive argument shows that

$$S_{m,n} = \frac{(n + 1)!}{(m + 1)(n - m)!},$$

and substitution then yields

$$\frac{(n - m)^{-1} S_{m+1,n}}{S_{m,n}} = \frac{(m + 1)}{(m + 2)}.$$

Broad did not give the mathematical details, and for completeness a proof is given in the appendix at the end of this paper.

The finite rule of succession has several important philosophical consequences:

(1) It eliminates a variety of possible concerns about the occurrence of the infinite in the Laplacean analysis (e.g., Kneale 1949, p. 205): attention is focused on a finite segment of trials, rather than a hypothetical infinite sequence.
(2) The frequency, propensity, or objective chance p that occurs in the integration is replaced by the fraction of successes; thus a purely personalist or subjective analysis becomes possible, and objections to "probabilities of probabilities" and "unknown probabilities" (e.g., Keynes 1921, pp. 372–75) are eliminated.
(3) It extends the domain of applicability of the rule to forms of enumerative induction not previously covered.

An important consequence of Broad's analysis was the remark that the probability of a universal generalization – i.e., that all n balls in the urn are black, given that the first m were – will be quite small unless m is large relative to n (the exact probability is $(m + 1)/(n + 1)$). This was not a novel observation, but it was viewed at the time as a serious setback to the Laplacean program of justifying induction probabilistically, and was an important impetus for the early work of Jeffreys and Wrinch (1919). This question will be discussed in the final sections of the paper.

Historical Note. Although Broad is often credited with the finite rule of succession (e.g., by von Wright 1957; Jeffreys 1961; Good 1965, p. 18), he does not specifically claim priority in its derivation, and in fact it had been independently discovered several times prior to Broad's 1918 paper. The first of these was in 1799, in a paper by the Swiss mathematicians Pierre Prevost (1751–1839) and Simon L'Huilier (1750–1840). Both were interested in the philosophical implications of probability and wrote several papers on the subject in collaboration; see generally Todhunter (1865, pp. 453–463).[3]

As Prevost and L'Huilier state the problem,

Soit une urne contenant un nombre n de billets; on a tiré $p + q$ billets, dont p sont blancs and q non-blancs (que j'appellerai noirs). On demande les probabilités que les billets blancs et les billets noirs de l'urne étoient des nombres donnés, dans la supposition qu'à chaque tirage on n'a pas remis dans l'urne le billet tiré.

Thus, Prevost and L'Huilier consider the more general case of p successes and q failures in $p + q = m$ trials, and derive the posterior probabilities for different constitutions of the urn. The law of succession is then derived as a consequence, with the result that the probability of a success on the next trial is $(p + 1)/(m + 2)$.

The result was later independently derived by Ostrogradskii (1848), as well as "a mere diocesan" (Keynes 1921, p. 179), Bishop Charles Terrot of Edinburgh, whose work (Terrot 1853) is mentioned by Boole in his *Investigation of the Laws of Thought* (1854). These early derivations are not without interest, and are discussed in the mathematical appendix at the end of this paper.

The result is also noted by the indefatigable Todhunter, who reports the work of Prevost and L'Huilier in his famous *History of the Mathematical Theory of Probability* (1865, pp. 454–57). Todhunter observes that the crucial sum may be readily evaluated by the use of the binomial theorem, remarks the identity of the answer with the Laplacean one, and comments that "the coincidence of the results obtained on the two different hypotheses is remarkable".[4]

3. FINITE EXCHANGEABILITY AND THE RULE OF SUCCESSION

Although the Prevost-L'Huilier and later proofs of the finite rule of succession are not difficult, they leave unexplained this "remarkable coincidence". It turns out that there is a very different approach, involving the concept of exchangeability, which clarifies why the finite and infinite rules agree.

Let $X_1, X_2, \ldots, X_n$ denote a sequence of exchangeable random variables taking on the values 0 and 1. By definition this means that the probability distribution of the random variables is invariant under permutations; i.e.,

$$P[X_1 = e_1, X_2 = e_2, \ldots, X_n = e_n]$$
$$= P[X_1 = e_{\sigma(1)}, X_2 = e_{\sigma(2)}, \ldots, X_n = e_{\sigma(n)}],$$

41

for all possible sequences $e_1, e_2, \ldots, e_n(e_i = 0$ or $1)$, and permutations σ of $\{1, 2, \ldots, n\}$. There is a simple representation for such sequences. If $S_n = X_1 + X_2 + \cdots + X_n$, then the events $\{S_n = k\}$ form a partition, i.e., they are mutually exclusive and exhaustive (they are disjoint and one of them must occur). Thus, by the so-called theorem on total probability, one may write

$$P[X_1 = e_1, X_2 = e_2, \ldots, X_n = e_n]$$
$$= \sum_{k=0}^{n} P[X_1 = e_1, X_2 = e_2, \ldots, X_n = e_n | S_n = k] P[S_n = k].$$

By the definition of exchangeability, the conditional probabilities $P[X_1 = e_1, X_2 = e_2, \ldots, X_n = e_n | S_n = k]$ assign equal probabilities to the $_nC_k$ sequences of k 1s and $n - k$ 0s. This corresponds to drawing at random all n balls out of an urn containing k 1s and $n - k$ 0s, i.e., it is a *hypergeometric probability* which we will denote $H_{n,k}$. Let $p_k = P[S_n = k]$. The sequence $p_0, p_1, \ldots, p_n$ specifies the probabilities that the sequence $X_1, X_2, \ldots, X_n$ will contain $0, 1, \ldots$ or n 1s, respectively. In this notation,

$$P = \sum_{k=0}^{n} p_k H_{n,k}.$$

That is, the exchangeable probability P may be viewed as a mixture of the hypergeometric probabilities $H_{n,k}$, using the p_k. If one were to arrange $n + 1$ urns $U_0, U_1, \ldots, U_n$, with urn U_k containing k 1s and $n - k$ 0s, pick an urn U_k with probability p_k, and then draw all n balls at random out of the urn, the resulting probability distribution on sequences of length n would be identical with the original probability assignment P.

This simple, but very useful result, is the *finite de Finetti representation theorem*. Note the following basic properties of the representation:

FE1. The $H_{n,k}$ are independent of P; P only enters into the representation via the p_k.

FE2. The representation is unique: if $P = \sum p_k H_{n,k} = \sum q_k H_{n,k}$, then $p_k = q_k$, all k.

FE3. The probability distribution on sequences of length n, arising from any mixture $\sum p_k H_{n,k}$, is exchangeable. (The term mixture means that the p_k are arbitrary numbers satisfying $0 \leq p_k \leq 1$ and $p_0 + p_1 + \cdots + p_n = 1$.)

In honor of those who were the first to study it, let us call the sequence generated by picking one of the $n + 1$ urns U_k at random, and then drawing all n balls out of the urn at random, the *Prevost–L'Huilier process*, denoted for short as PL_n. The Prevost–L'Huilier process is a special example of a finite exchangeable sequence, with the $p_k = P[S_n = k] =: 1/(n+1)$ uniform. It is a consequence of FE1 that an exchangeable sequence is uniquely determined once the values $p_k = P[S_n = k]$ are specified.

Now we are ready to explain the strange coincidence of the rules of succession for the Prevost–L'Huilier process PL_n and the Bayes-Laplace process BL_∞, which is generated by picking p uniformly from the unit interval $[0, 1]$ and then tossing a p-coin infinitely often. The Bayes–Laplace process $X_1, X_2, X_3, \ldots$ is an infinitely exchangeable sequence; i.e., for any $n \geq 1$, the initial segment of the process $X_1, X_2, \ldots, X_n$ is exchangeable in the sense defined above. Thus it has some finite de Finetti representation $\sum p_k H_{n,k}$. But, the Bayes–Laplace process BL_∞ has the property that $p_k = P[S_n = k] = 1/(n+1)$, just as does the Prevost–L'Huilier process PL_n. Since they are both exchangeable, and since their mixing measures coincide, they are *identical*. That is,

> the initial segment $X_1, X_2, \ldots, X_n$ of the Bayes–Laplace process BL_∞ is stochastically identical to the Prevost–L'Huilier process PL_n.

Now it is clear why the rules of succession for the two processes coincide: they are actually the *same* process (up to stage n)! Not only do their rules of succession coincide, but every other probabilistic aspect as well. Although the two processes were generated by two distinct stochastic mechanisms, the resulting distributions are identical.

In retrospect, this is obvious: if we are given the initial probabilities $P[X_1 = e_1]$, and the rules of succession at each stage, it is possible to express the probabilities $P[X_1 = e_1, X_2 = e_2, \ldots, X_n = e_n]$ in terms of these quantities. For example, for both PL_4 and BL_∞,

$$P[X_1 = 1, X_2 = 0, X_3 = 1, X_4 = 1] = (1/2)(1/3)(2/4)(3/5) = 1/20.$$

Thus, if the initial probabilities and succession probabilities of two processes coincide, the processes are the same. For those allergic to exchangeability arguments of the type given above, this gives an alternative way of deriving the identity of PL_n and the initial nth segment of BL_∞ once their rules of succession have been shown to coincide.

Observing the identity of the two processes has the advantage that most properties of PL_n may be immediately and easily deduced. For example, consider the following question:

> Given a sequence of total length N, what is the probability that if the first n outcomes are all black, then the remaining $N - n$ outcomes will also be all black?

That is, how much evidence does the first n outcomes being black provide toward the universal generalization that all outcomes are black? Doing this directly (as Broad did) is elementary but messy, involving the usual sums. Far easier, however, is the observation that

$$
\begin{aligned}
P[S_N = N | S_n = n] &= P[S_N = N \text{ and } S_n = n]/P[S_n = n] \\
&= P[S_N = N]/P[S_n = n] \\
&= \frac{1}{(N+1)} \bigg/ \frac{1}{(n+1)} \\
&= \frac{(n+1)}{N+1},
\end{aligned}
$$

which is the answer Broad derives, and which coincides (as it must) with the result for the Bayes–Laplace process (Laplace, *Théorie analytique*, p. 402; De Morgan 1838, p. 64).

How satisfactory an explanation of enumerative induction does the rule of succession provide? What are its limitations? Can these be eliminated? Broad's analysis came at the end of a century and a half of discussion and debate. It marks the end of an era, for in a few years the contributions of Keynes, Johnson, Ramsey, and de Finetti were to irretrievably change the way in which the problem was cast. The next three sections discuss some of the highlights of the preceding debate from Bayes to Broad. Those readers not interested in this previous history may turn directly to Section 7, where the emphasis shifts to the philosophical analysis and mathematical evolution of the rule.

4. WHEN AND WHY DID BAYES PROVE BAYES'S THEOREM?

Hume first stated the problem of induction; Bayes first advanced a solution to it. The chronological link between these two events is much closer than is usually recognized.

Like James Bernoulli before him, the Reverend Thomas Bayes perished before he was published. At some time prior to his death on 17 April 1761, Bayes wrote his famous 'Essay Towards Solving a Problem in the Doctrine of Chances', published posthumously by his friend Richard Price in 1764. Although Bayes's introduction to his essay has not survived, Price tells us that Bayes came to have doubts as to the validity of the postulate adopted by him in the solution of the problem. As Price puts it, Bayes "afterwards considered, that the postulate on which he had argued might not perhaps be looked upon by all as reasonable; and therefore he chose to lay down in another form the *proposition* in which he thought the solution of the problem is contained, and in a *scholium* to subjoin the reasons why he thought so, rather than to take into his mathematical reasoning any thing that might admit dispute."

For this reason some latter commentators have assumed that Bayes delayed publication of his results because of such doubts (e.g., Fisher 1973, pp. 9–10). How long did Bayes meditate on his solution? Surprisingly, there is evidence that suggests that Bayes may have arrived at at least the basic results in his essay some fifteen years prior to his death.

The first piece of evidence in question is a passage from David Hartley's *Observations on Man*, published in 1749. After discussing the law of large numbers for binomial trials given by De Moivre, Hartley states

An ingenious Friend has communicated to me a Solution of the inverse Problem, in which he has shewn what the Expectation is, when an Event has happened p times, and failed q times, that the original Ratio of the Causes for the Happening or Failing of an Event should deviate in any given Degree from that of p to q. And it appears from this Solution, that where the Number of Trials is very great, the Deviation must be inconsiderable: Which shews that we may hope to determine the Proportions, and, by degrees, the whole Nature, of unknown Causes, by a sufficient Observation of their Effects. (Hartley 1749, p. 339)

If Hartley's ingenious friend were Bayes, this would mean that Bayes had arrived at his basic results no later than 1749, and probably somewhat earlier. The identity of the two is not only a natural conjecture, it is supported by the internal evidence of Hartley's own statement: the terminology used by Hartley is identical to that employed by Bayes, who refers in his essay to an "event . . . happening p times, and failing q times . . . ". (Ingenious, moreover, was a word which came readily to mind when thinking of Bayes. Price, for example, calls Bayes "one of the most ingenious men I ever knew" (Price 1758, p. 248), and Laplace refers to Bayes's method as "très ingénieuse" (Laplace 1814, p. cxlviii).)

If Bayes did suppress his result for some 15 years, his diffidence in publication might well explain the anonymous nature of Hartley's reference. Since Bayes and Hartley were both members of the Royal Society and dissenters, they may well have known each other, although there is no direct evidence that they actually did. It is of course possible that Hartley's "ingenious friend" was someone other than Bayes, but absent Hartley's direct statement to this effect or clear evidence that Bayes's work had been independently duplicated, it is hard to credit this hypothesis.[5]

Very recently a new piece of evidence has come to light that seems decisive in favor of Hartley's reference to Bayes. Dr. A. I. Dale has discovered a passage in an early notebook of Bayes giving a proof of one of the rules in Bayes's essay (Dale 1986). Although the entry is undated, it is preceded by one dated July 4, 1746, and succeeded by one dated December 31, 1749. It is thus clear that at some point in the period 1746–1749 Bayes had derived at least some of his basic results, and the coincidence with the date of Hartley's book (1749) seems too striking to be coincidental.

What event in the period 1746 to 1749 led Bayes to investigate a problem that, in the words of his friend Richard Price, must be "considered by any one who would give a clear account of the strength of *analogical* or *inductive reasoning*"? Thus put, an obvious answer suggests itself. In 1748 David Hume had published his *Enquiries Concerning Human Understanding*, containing a clear and succinct statement of his famous problem of induction. Hume had laid down the challenge: "Let any one try to account for this operation of the mind upon any of the received systems of philosophy, and he will be sensible of the difficulty" (*Enquiry*, p. 59). Bayes may have answered it within a year.

Bayes's paper, however, had little immediate, direct influence and it is through Laplace that the techniques of inverse probability became widely known. A decade after the appearance of Bayes's essay, Laplace wrote the first of a series of papers in which he, apparently independently of Bayes, presented his solution to the problem of causes, in the form that was to gain widespread acceptance (Laplace 1774).[6] His older mentor Condorcet, recognizing the importance of Laplace's contribution to the inductive problem, rushed it into print. "The problem of finding the probability of the occurrence of an event, given only that it has occurred a number of times in the past, is the most fundamental in the calculus of probabilities, argued the assistant secretary [Condorcet], underlining the significance of Laplace's paper in the preface to the sixth volume of the *Mémoires des savants étrangers*" (Baker 1975, pp. 168–69). Hume's impact had been felt on the Continent as well.[7]

Laplace's own statement of the probabilistic solution of the problem of induction appears in the *Essai philosophique*. The example he provided is notorious:

Thus we find that an event having occurred successively any number of times, the probability that it will happen again the next time is equal to this number increased by unity divided by the same number, increased by two units. Placing the most ancient epoch of history at five thousand years ago, or at 1826213 days, and the sun having risen constantly in the interval at each revolution of twenty-four hours, it is a bet of 1826214 to one that it will rise again tomorrow. [Laplace, *Essai*, p. xvii]

It is said that Laplace was ready to bet 1,826,214 to 1 in favor of regular habits of the sun, and we should be in a position to better the odds since regular service has followed for another century. A historical study would be necessary to appreciate what Laplace had in mind and to understand his intentions. (Feller 1968, p. 124)

Laplace has perhaps received more ridicule for this statement than for any other. Yet Feller, despite his general lack of sympathy for the Bayesian position, had too much respect for Laplace to dismiss his famous calculation unexamined. Let us attempt the study Feller suggests.

To begin with, it is important to realize that the example of the rising of the sun does not originate with Laplace. It goes back to Hume (at least), who in his *Treatise* of 1739 asserted: "One wou'd appear ridiculous, who wou'd say, that 'tis only probable the sun will rise to-morrow, or that all men myst dye; tho'.'tis plain we have no further assurance of these facts, than what experience affords us" (*Treatise*, p. 124). As we shall see, the example of the rising of the sun as a touchstone of inductive inference is a common thread through much of the later literature on the subject.

In denying that inferences such as the rising of the sun are merely probable, Hume was arguing that there are degrees of knowledge which, while not demonstratively certain, exceed all probability. This is a recurrent idea, which can also be found, for example, in Cardinal Newman's *Grammar of Assent*. Price, to contradict Hume, turns to this example in his appendix to Bayes's essay:

Let us imagine to ourselves the case of a person just brought forth into this world, and left to collect from his observation of the order and course of events what powers and causes take place in it. The Sun would, probably, be the first object that would engage his attention; but after losing it the first night he would be entirely ignorant whether he should ever see it again. He would therefore be in the condition of a person making a first experiment about an event entirely unknown to him. But let him see a second appearance or one return of the Sun, and an expectation would be raised in him of a second *return*, and he might know that there was an odds of 3 to 1 for *some* probability of this. This odds would increase, as before represented, with the number of returns to which he was witness. But no finite number of returns would be sufficient to produce absolute or physical certainty. For let it be supposed that he has seen it return at regular and stated intervals a million of times. The conclusions this would warrant would be such as follow. There would be the odds of the millioneth power of 2, to one, that it was likely that it would return again at the end of the usual interval.

This is *not* Laplace's rule of succession, but rather a calculation of the posterior probability that the unknown chance p of the sun's rising exceeds 1/2, i.e.,

$$P[p > 1/2] = \int_{1/2}^{1} p^{n-1} dp \Big/ \int_{0}^{1} p^{n-1} dp$$
$$= 1 - (1/2)^n = (2^n - 1)/2^n.$$

i.e., odds of 2^n to 1. (Note Price uses an exponent of $n - 1$, since he considers the first trial to merely inform us that the event is possible; see Pearson 1978, pp. 368–69.)[8]

Although Price was a lifelong philosophical opponent of Hume, he read Hume carefully, and it is clear that his discussion of Hume's example was intended to rebut Hume's contention that "many arguments from causation exceed probability, and may be receiv'd as a superior kind of evidence . . . . which are entirely free from doubt and uncertainty." Indeed, not only does Price address Hume's example, but he goes on to stress that "instead of proving that events will always happen agreeably to [uniform experience], there will always be reason against this conclusion."[9]

But even if one concedes that our knowledge of future events such as the rising of the sun only admit of probability, there is a leap of faith in Price's argument. Price began his analysis by first considering "a solid or die of whose number of sides and constitution we know nothing; and that we are to judge of these from experiments made in throwing it," later explaining that he "made these observations chiefly because they are all strictly applicable to the events and appearances of nature." Condorcet, in his *Essai*, accepts this nexus without reservation:

Ainsi le motif de croire que sur dix millions de boules blanches mêlées avec une noire, ce ne sera point la noire que je tirerai du premier coup, est de la même nature que le motif de croire que le Soleil ne manquera pas de se lever demain, & ces deux opinions ne different entr'elles que par le plus ou le moins de probabilité. (Condorcet 1785, p. xi)

This was a sweeping claim, and it did not pass unchallenged. Prevost and L'Huilier, in a philosophical essay accompanying their paper on the finite rule of succession, soon took issue with Condorcet, arguing

La persuasion analogique qu'éprouve tout homme, de voir se répéter un événement naturel (tel que le lever du soleil), est d'un genre différent de la persuasion représentée par une fraction dans la théorie des probabilités. Celle-ci peut lui être ajoutée, mais l'une peut exister sans l'autre. Elles dépendent de deux orders de facultés différens. Un enfant, un animal éprouve la première, & ne forme aucun calcul explicite, ni même implicite: il n'y a aucune dépendance nécessaire entre ces deux persuasions. Celle que le calcul apprécié est raisonné, & même, jusqu'à un certain point, artificielle. L'autre est d'instinct & naturelle. Elle dépend de quelques facultés intellectuelles dont l'analyse n'est pas facile, & probablement en très-grande partie du principe de la liaison des idées. (Prevost and L'Huilier 1799a, p. 15)

This is one of the earliest arguments urging the distinction between induction ("la persuasion analogique") and probability ("une fraction dans la théorie des probabilités"), and it presages a debate that continued unabated through the next century. (For the possible influence of Prevost and L'Huilier on Mill, via Dugald Stewart, see Strong 1978, p. 35). Bertrand, for example, writing nearly a hundred years later in his distinctively acerbic French prose, singles out the same passage from Condorcet for criticism:

L'assimilation n'est pas permise: l'une des probabilités est objective, l'autre subjective. La probabilité de tirer la boule noire du premier coup est 1/10 000 000, ni plus ni moins. Quiconque l'évalue autrement se trompe. La probabilité pour que le Soleil se lève varie d'un esprit à l'autre. Un philosophe peut, sans être fou, annoncer sur la foi d'une fausse science que le Soleil va bientôt s'éteindre; il est dans son droit comme Condorcet dans le sien; tous deux l'excéderaient en accusant d'erreur ceux qui pensent autrement. (Bertrand 1907, p. xix.)

Many other examples could be adduced.[10] What is striking in many of these discussions is the virtual lack of serious argument. Positions are staked out, but there is often surprisingly little in the way of genuine analysis or critical discussion. (One exception is Bertrand 1907, pp. 173–74).

A common position was that such inductive inferences, even if "probable", could not be quantified – that what was in question was a species of *philosophical* probability, rather than *mathematical* probability. Strong (1978, p. 207, n. 5) cites an early example of this distinction in a rare work of K. H. Frömmichen of 1773. It will be apparent by now that the date is "no accident"; by this time the claims of probability in natural philosophy were beginning to provoke dissent.

Such considerations were not, however, foreign to Laplace. His rule of succession is an instrument for "pure inductions," or "eduction" as W. E. Johnson later termed them. That Laplace was not under the illusion that "hypothetical inductions" could also be so described is clear from the penultimate chapter of the *Essai philosophique*, "Concerning the various means of approaching certainty." At the end Laplace cautions:

It is almost always impossible to submit to calculus the probability of the results obtained by these various means; this is true likewise for historical facts. But the totality of the phenomena explained, or of the testimonies, is sometimes such that without being able to appreciate the probability we cannot reasonably permit ourselves any doubt in regard to them. In the other cases it is prudent to admit them only with great reserve.

De Morgan, too, later cautioned that "in the language of many, induction is used in a sense very different from its original and logical one. . . . What is now called induction, meaning the discovery of laws from instances, and higher laws from lower ones, is beyond the province of formal logic" (De Morgan 1847, p. 215). (Note from the title of his book that De Morgan includes probability within that province.)

Thus, when Laplace made his notorious remark in the *Essai philosophique*, he was writing against a background of 75 years of debate and discussion about inductive inference throughout which the example of the rising of the sun runs as a common and recurrent thread. In his dry style, Laplace omits virtually all reference to this previous debate.

How seriously did Laplace view the calculation itself? Certainly much less so than is usually implied. All too often it is cited out of context, for after the passage quoted, Laplace went on to immediately add:

But this number is incomparably greater for him who, recognizing in the totality of phenomena the regulatory principle of days and seasons ["connaissant par l'ensemble des phénomènes le principe régulateur des jours et des saisons"], sees that nothing at the present moment can arrest the course of it.

The point is clear: the calculation only establishes the probability that flows from the mere repetition of events.[11] And while Laplace did not belabor the point, he was far from the only one to make it. Price too had cautioned that "it should be carefully remembered that these deductions suppose a previous total ignorance of nature", and his fanciful narrative is clearly intended to stress the artificial nature of the assumption. When Quetelet gives a similar analysis for the rising of the tides, it is for someone who has never seen them before. The English logician and mathematician Augustus De Morgan, who played an important role in disseminating Laplace's work during the nineteenth century, also stressed the point, terming the rule of succession "the rule of probability of a *pure induction*", and adds that "the probabilities shown by the above rules are merely *minima* which may be augmented by other sources of knowledge" (De Morgan 1847, p. 215).

6. THE GREAT JEVONIAN CONTROVERSY

This then was the Laplacean contribution to the probabilistic analysis of enumerative induction. How did it fare during the nineteenth century?

The English logician William Stanley Jevons is often portrayed as the first important philosopher of science to systematically link probability and induction (Keynes 1921, p. 273; Madden 1960, p. 233; Heath 1967, p. 261; Laudan 1973). Indeed, in Laudan (1973), the history of the subject revolves around Jevons: why did inductive logicians and philosophers of science before Jevons spurn probability; why did another half-century have to pass after Jevons before the link between probability and induction was taken seriously? Laudan considers these issues, centering his discussion on Jevons's arguments in favor of the link, and its criticisms by the English logician John Venn.

Laudan's analysis is largely vitiated, however, by a surprising chronological error: he presents Venn's criticisms as – and apparently believes them to be – an attack on Jevons, despite the fact that the 1st edition of Venn's *Logic of Chance* appeared in 1866, eight years prior to the appearance of the 1st edition of Jevons's *Principles of Science* (1874). Although it is true that Venn made extensive revisions in the 2nd (1876) and 3rd (1888) editions of the *Logic*, the vital chapter on 'Induction and its Connection with Probability' goes back to the 1st, and while the 1888 edition of the *Logic* (which Laudan quotes) does refer on several occasions to Jevons's *Principles*, it does so only briefly: despite several passages where the wording has been recast, new material added, or the text shortened, the basic thrust and content of the chapter remains that of the 1st edition.

But if Venn was not, at least initially, directing his fire against Jevons, who then? The answer is clearly the English mathematician and logician Augustus De Morgan. De Morgan was Laplace's most enthusiastic English advocate, the author of no fewer than three works during the decade 1838–1847 intended to popularize and spread the Laplacean approach to probability.[12] Indeed, De Morgan's *Formal Logic* of 1847 was the first English language textbook on logic to break with tradition by presenting probability as a branch of formal logic, a precedent followed by Boole several years later in the latter's *Investigation of the Laws of Thought* of 1854. Venn explicitly singles De Morgan out, saying that he would have felt no need to write *The Logic of Chance*, given De Morgan's writings on probability, save that he differed from De Morgan in too fundamental a way (Venn 1888, p. ix). (Jevons in fact was a student of De Morgan's, and it was from De Morgan that he learned probability theory.)

Jevons was thus not alone. The probabilistic basis of at least some forms of induction had been advocated prior to Jevons by Condorcet, Laplace, Lacroix, Quetelet, Herschel, and De Morgan, and after Jevons by Peirce, Pearson, and Poincaré. Jevons was neither the first to argue the connection, nor the first philosopher of science or inductive logician to do so, but among this latter tribe he was admittedly one of the few to do so. As Venn testifies,

So much then for the opinion which tends to regard pure Induction as a subdivision of Probability. By the majority of philosophical and logical writers a widely different view has of course been entertained. They are mostly disposed to distinguish these sciences very sharply from, not to say to contrast them with, one another; the one being accepted as philosophical or logical, and the other rejected as mathematical. This may without offence be termed the popular prejudice against Probability. (Venn 1888, pp. 208–209)

"Why did we have to wait for Stanley Jevons, and C. S. Peirce, writing in the 1870s, rather than Hume in the 1740s or Mill in the 1840s, to find someone systematically arguing that inductive logic is based on probability theory?" (Laudan 1973, p. 429). For Hume, there is a simple answer: the necessary apparatus of inverse probability did not exist when he wrote his *Treatise* and *Enquiries*. As discussed earlier, both Bayes and Laplace were aware of the relevance of their contributions to the questions addressed by Hume.

But what of the period after Laplace? Even if one takes 1814, the year of publication of the *Essai philosophique* as a point of departure, what happened in the 60 years that elapsed before the publication of Jevons's *Principles*? That De Morgan should embrace the Laplacean position on induction is not surprising; as we have noted, De Morgan was Laplace's staunchest English advocate and his writings on probability were in large part a deliberate effort to bring Laplace's work to the attention of the English public.

But why were there so few others in the English philosophical community to embrace the Laplacean position? Here the answer is not complimentary to English philosophy: the mathematical prerequisites were such as to exclude most writers on the subject. On this we have the testimony of Venn himself, the archcritic of Laplacean probability:

The opinion that Probability, instead of being a branch of the general science of evidence which happens to make much use of mathematics, *is* a portion of mathematics, erroneous as it is, has yet been very disadvantageous to the science in several ways. Students of Philosophy in general have thence conceived a prejudice against Probability, which has for the most part deterred them from examining it. As soon as a subject comes to be considered 'mathematical' its claims seem generally, by the mass of readers, to be either on the one hand scouted or at least courteously rejected, or on the other to be blindly accepted with all their assumed consequences. Of impartial and liberal criticism it obtains little or nothing. (Venn 1888, p. vii)

Interestingly, Venn sees as the most unfortunate result of this neglect the loss for probability rather than the loss for philosophy: "The consequences of this state of things have been, I think, disastrous to the students themselves of Probability. No science can safely be abandoned entirely to its own devotees." Probability is too important to be left to the mathematicians.

This then, was the background against which Jevons wrote. In truth, there is little new in Jevons, but despite his many weaknesses, he represents a clear and succinct statement of the Laplacean position. Nevertheless, nearly half a century was to pass before Jevons's program was to be pushed forward by philosophers such as Johnson, Broad, Keynes, and Ramsey.

This hiatus, however, is not surprising. During the decades immediately following the appearance of Jevons's book, epistemic probability was pre-occupied. The two-pronged assault of Boole (on the logical front) and Venn (on the empirical front) had called into serious question the very foundations of the Laplacean edifice. Epistemic probability did not go under during this period (Zabell 1989), but it did have to put its foundational house in order before it could contemplate expanding its horizons. After the contributions of Johnson, Keynes, Ramsey, and de Finetti this became possible.

Although the old Laplacean approach to probability eventually died out, epistemic probability arose transfigured from its ashes. While some continued to defend the principle of indifference – indeed, some still do – the key step in this metamorphosis was the abandonment of uniform priors and, on the inductive front, any attempt at a unique quantitative explanation of inductive inference.

A complete account of this transformation has never been written, and would go far beyond the compass of the present study. But limiting our attention to charting the vicissitudes of the rule of succession throughout the following period provides the opportunity for a case study, highlighting in a microcosm many of the arguments and issues that arose in the broader debate.

7. DEATH AND TRANSFIGURATION

As the statistician R. A. Fisher once noted, the rule of succession is a mathematical consequence of certain assumptions, and its application to concrete examples can only be faulted when the examples fail to satisfy the presuppositions. Those presuppositions involve two distinct types of issues. At the level of balls in an urn, there is the assumption that the possible urn compositions are equally likely, i.e., the principle of indifference. And at the level of applying the resulting mathematics to the real world, there is the question of the relevance of the urn model. The attacks on the rule of succession involved both of these points.

7.1. The Principle of Indifference

The Achilles' heel of the rule of succession lies in its appeal to the principle of indifference. It assumes that all possible ratios are equally likely, and that in particular, on any single trial, the probability of an event "concerning the probability of which we absolutely know nothing antecedently to any trials concerning it" (Bayes 1764, p. 143), is 1/2. For example, in the analysis of the rising of the sun, it is assumed to be equally likely that the sun will or will not rise.

Apart from ridicule, this position was subjected to a number of telling criticisms, particularly by Boole (1854) and von Kries (1886), and a large number of now-standard paradoxes and counterexamples were adduced (for von Kries, see Kamlah 1983 and 1987). A common response to many of these examples is to point to the requirement of the absence of prior knowledge about the event in question, and argue that it is violated. The fatal flaw in all such defenses is that understanding the very words employed in describing an event necessarily implies some knowledge about that event. Thus, as Keynes notes, in Jevons's notorious example of the proposition "a platythliptic coefficient is positive", the force of the example derives from our entire ignorance of the meaning of the adjective "platythliptic" (Keynes 1921, p. 42, n. 2). Nevertheless, the example is defective, given we do possess considerable knowledge about the words "coefficient" and "positive". Keynes is not being sarcastic, but merely pursuing the argument to its logical conclusion when he asks whether Jevons would "maintain that there is any sense in saying that for those who know no Arabic the probability of every statement expressed in Arabic is even?" (Keynes 1921, p. 43).

Even at the syntactic level, it is easy to construct contradictory probability assignments using the principle of indifference whenever a complex proposition can be decomposed into simpler ones. If Wittgenstein's early program of logical atomism had been successful, then logical probability would be possible, but the failure of the former dooms all attempts to construct the latter. Lacking an ultimate language in one-to-one correspondence with reality, Carnapian programs retain an ultimate element of subjectivism, both in their choice of language and the assumption that a given partition consists of equiprobable elements.

For essentially such reasons, von Kries and others fell back on what was called the principle of *cogent* reason: alternatives are equally probable when we possess knowledge about them, but that knowledge is equally distributed or symmetrical among the alternatives. This was, in fact, the actual Laplacean position: "la probabilité est relative en partie à cette ignorance, en partie à nos connaissances" (Laplace, *Essai*, p. viii). The formulation of the principle of cogent reason, of course, is not without its own problems, and its most satisfactory statements verge on the tautologous. It was, however, a half-way house on the road to the only truly satisfactory formulation: alternatives are equally probable when we judge them to be so. Assignments of equiprobability can only originate as primitives of the system, inputs that are *given*, rather than logical consequences of the syntax of language. Ellis was entirely correct: *ex nihilo nihil.*

7.2. The Urn of Nature

The valid application of the rule of succession presupposes, as Boole notes, the aptness of the analogy between drawing balls from an urn – the *urn of nature*, as it was later called – and observing an event (Boole 1854, p. 369). As Jevons put it, "nature is to us like an infinite ballot-box, the contents of which are being continually drawn, ball after ball, and exhibited to us. Science is but the careful observation of the succession in which balls of various character present themselves . . ." (p. 150).

The origins of the urn of nature are perhaps to be found in James Bernoulli's *Ars conjectandi*. This was a key moment in the history of probability, when the domain of applicability of the theory was dramatically broadened to include physical, biological, and social phenomena far beyond the simple applications to games of chance originally envisaged. But lacking a suitable frequentist or epistemic foundation for probability, Bernoulli was forced to employ the Procrustean bed of equally likely cases: "the numbers of cases in which the same events, with similar circumstances prevailing, are able to happen and not to happen later on". In attempting to apply the doctrine of chances to questions of meteorology, human mortality, and competitive skill, Bernoulli saw the difficulty as one of enumerating these equipossible cases; for example, "the innumerable cases of mutations to which the air is daily exposed", or "the number of diseases". Who, Bernoulli asks, "has well enough examined the nature of the human mind or the amazing structure of our body so that in games which depend wholly or in part on the acumen of the former or the agility of the latter, he could dare to determine the cases in which this player or that can win or lose?" This is the origin of the urn of nature.

What is remarkable about these passages in the *Ars conjectandi* is the almost casual way in which Bernoulli passes from equally likely cases for games of chance to what is essentially a primitive form of propensity theory for physical, biological, and social phenomena. Price, too, began his analysis by first considering "a solid or die of whose number of sides and constitution we know nothing; and that we are to judge of these from experiments made in throwing it", later explaining that he "made these observations chiefly because they are all strictly applicable to the events and appearances of nature".

The aptness of this analogy between tossing a die, or drawing a ball from an urn, is one of the great points in the later debate. Some, like Comte, utterly rejected the application of probability theory outside its original narrow domain, referring contemptuously to Laplace's work as embodying a "philosophical aberration". Others might accept a probabilistic description of sex at birth, or suicide, or weather, but questioned the appropriateness of the analogy in cases such as the rising of the sun, or the movement of the tides.

Thus for enumerative induction the key question became: why and in what way can the relevant observations be viewed as drawings from an urn?

7.3. W. E. Johnson's Rule of Succession

In 1924 the English philosopher and logician William Ernest Johnson published the third and final volume of his *Logic*. In an appendix on "eduction" (i.e., inductive inference from particulars to particulars), Johnson derived a new rule of succession which met both of these basic objections. First, "instead of for two, my theorem holds for α alternatives, primarily postulated as equiprobable" (Johnson 1932, p. 418). Thus the principle of indifference *for alternatives* was exorcised, and the rule extended to cases of multinomial sampling. Although Johnson's form of the rule is sometimes viewed as a straightforward generalization of the original, it will now be appreciated why the generalization was crucial. (Although a proposition and its negation might not be judged equiprobable, the proposition might be one of a spectrum of possibilities which were.)

The mere multinomial generalization, however, had already been discussed by Laplace and De Morgan.[13] But in its derivation Johnson introduced a new and important concept: *exchangeability*. Johnson assumed that "each of the different orders in which a given proportion $m_1 : m_2 : \cdots : m_\alpha$ for M instances may be presented is as likely as any other, what ever may have been the previously known orders". Johnson termed this the "Permutation-Postulate". Its importance is that *it is no longer necessary to refer to the urn of nature*. To what extent is observing instances like drawing balls from an urn? Answer: to the extent that the instances are judged exchangeable. Venn and others had adduced examples where the rule of succession was clearly inappropriate and rightly argued that some additional assumption, other than mere repetition of instances, was necessary for valid inductive inference. From time to time various names for such a principle have been advanced: Mill's Uniformity of Nature; Keynes's Principle of Limited Variety; Goodman's "projectibility". It was Johnson's achievement to have realized both that "the calculus of probability does not enable us to infer any probability-value unless we have some probabilities or probability relations given" (Johnson 1924, p. 182); and that the vague, verbal formulations of his predecessors could be captured in the mathematically precise formulation of exchangeability.[14]

But the rule of succession does not follow from the assumption of exchangeability alone. As we have already seen in Section 3, an assumption must be made about the probability of the different urn compositions. Johnson called the assumption he employed the *combination postulate*: In a total of M instances, any proportion, say $m_1 : m_2 : \ldots : m_\alpha$, where $m_1 + m_2 + \cdots + m_\alpha = M$, is as likely as any other, prior to any knowledge of the occurrences in question (Johnson 1924, p. 183). This is the multinomial generalization of the Bayes–Laplace assumption that all proportions k/n are equally likely in the binomial case.

Given the permutation and combination postulates, Johnson was able by simple algebra to deduce the multinomial generalization of the rule of succession: $(m_i + 1)/(M + \alpha)$. Because of the setting, infinite trials never came into consideration, and thus this provided a multinomial generalization of the Prevost–L'Huilier/Broad result (although by a clever argument Johnson was able to avoid the problem of explicitly summing the relevant series).

Johnson's result thus coped with two of the three major defects in the rule of succession. If it went largely unappreciated, it was because it was soon superceded by other, more basic and fundamental advances.

7.4. W. E. Johnson's Sufficientness Postulate

The one remaining defect in the rule of succession, as derived by Johnson, was its assumption of the combination postulate. Although Johnson made no appeal to the principle of indifference, the justification for the combination postulate seemed problematical. Johnson himself recognized this, for he soon proposed another, more general postulate, the "sufficientness postulate": the probability of a given type, conditional on n previous outcomes, only depends on how many instances of the type in question occurred, and not on how the other instances distributed themselves amongst the other types (Johnson 1932). Johnson was then able to show that the rule of succession in this case was $(m_i + k)/(M + k\alpha)$, where k can be any positive number. That is, assuming only the sufficientness postulate, a unique answer is no longer determined. This new rule is, of course, none other than Carnap's later "continuum of inductive methods".[15]

7.5. De Finetti and the Rule of Succession

The one final step that remained to be taken was the realization that it was unnecessary to make any further assumption beyond exchangeability. As de Finetti noted in his famous 1937 article, there is a general form of the rule of succession which holds true for an arbitrary finite exchangeable sequence. Namely, if $\omega_r^{(n)}$ denotes the probability of r successes in n trials, then the succession probability given r successes and s failures in $r + s = n$ trials is

$$\frac{r + 1}{n + 2 + (s + 1)(\omega_r^{(n+1)}/\omega_{r+1}^{(n+1)} - 1)},$$

(de Finetti 1937, p. 144). If $\omega_r^{(n+1)} = \omega_{r+1}^{(n+1)}$, then this reduces to the classical rule of succession. The condition is satisfied exactly if the classical uniformity assumption is made, or approximately in many cases for large n. Venn, in earlier editions of the *Logic of Chance*, had objected to the rule, adducing anti-inductive examples where past successes make future successes less likely rather than more; the de Finetti version of the rule of succession encompasses such situations.

De Finetti's analysis, appearing nearly two centuries after the appearance of Hume's *Treatise* in 1739, represents a watershed in the probabilistic analysis of induction. It abolishes all reference to the infinite, all reference to the principle of indifference, all reference to probabilities of probabilities, all reference to causation, all reference to principles of limited independent variety and other extraneous assumptions. In order to attack it, one must attack the formidable edifice of epistemic probability itself. Modern philosophy continues to ignore it at its own peril.

8. UNIVERSAL GENERALIZATIONS

In 1918 Broad had noted that if there are N balls in an urn, and all n in a random sample are black, then (under the usual equiprobability assumption), the probability that all the balls in the urn are black is $(n + 1)/(N + 2)$. If n is considerably smaller than N, this probability will also be small; i.e., unless a considerable segment of the sequence $X_1, X_2, \ldots, X_N$ has been observed, or a considerable number of balls drawn from the urn, or most ravens observed, the probability of the universal generalization will be small. This observation has been persistently viewed as an essentially fatal defect of this form of reducing induction to probability, going back at least to the neo-Kantian J. J. Fries in 1842.[16]

The assumption on which the calculation is based, that $P[S_N = k] = (N + 1)^{-1}, 0 \leq k \leq N$, is a *symmetry* assumption and, like many symmetry assumptions, contains "hidden baggage" not always initially apparent. In this case the hidden baggage lies surprisingly close to the surface; adopting the uniformity assumption $P[S_N = k] = 1/(N + 1)$ means assuming in particular that

$$P[S_N = 0] = P[S_N = N] = \frac{1}{(N + 1)},$$

i.e., that *the universal generalizations* $\{S_N = 0\}$ *and* $\{S_N = N\}$ *are a priori highly improbable* (if N is large). It is hardly surprising that hypotheses initially thought highly improbable should remain improbable unless a considerable fraction of the sequence has been observed.

If one deals with an infinitely exchangeable sequence, the problem becomes even more acute: taking the limit as $N \to \infty$ shows that for the Bayes–Laplace process the conditional probability of a universal generalization given n successes is always zero:

$$P[X_{n+1} = 1, X_{n+2} = 1, X_{n+3} = 1, \ldots, |S_n = n] = 0.$$

Once again, this is only surprising if one fails to understand what is being assumed. In the Bayes–Laplace process, the prior probability is zero that the unknown chance p equals one (i.e., $P[p = 1] = 0$), and thus the probability that the limiting frequency of 1s is one must necessarily also be zero.

There are two ways out of this dilemma for those who wish to conclude inductively that repeatedly confirmed universal hypotheses are a posteriori probable:

(1) *Stonewalling*. That is, argue that the initial intuition was in fact false; no matter how many successes have been observed does not warrant expecting arbitrarily long further strings of success. This is a very old defense, and it appears both in Price's appendix to Bayes's essay (p. 151), and Laplace's first paper on inverse probability (Laplace 1774).

De Morgan gives a good statement of the position:

[E]xperience can never, on sound principles, be held as foretelling all that is to come. The order of things, the laws of nature, and all those phrases by which we try to make the past command the future, will be understood by a person who admits the principles of which I treat as of limited application, not giving the highest degree of probability to more than a definite and limited continuance of those things which appear to us most stable. *No finite experience whatsoever can justify us in saying that the future shall coincide with the past in all time to come, or that there is any probability for such a conclusion.* (De Morgan, 1838, p. 128; emphasis De Morgan's)

Obviously such a position is open to the objection that some inductive inferences *are* of universal character, and it has been subjected to varying forms of ridicule. Keynes (1921, p. 383) notes that the German philosopher Bobeck calculated that the probability of the sun's rising every day for the next 4,000 years, given Laplace's datum that it has risen for the last 5,000 years or 1,826,213 days, is no more than 2/3. (Actually, using the formula above, the probability may be readily calculated to be $(1,826,213 + 1)/(3,287,183 + 1) = 0.56$.)

(2) *Different priors*. Edgeworth, in his review of Keynes's *Treatise*, remarks that "pure induction avails not without some finite initial probability in favour of the generalisation, obtained from some other source than the instances examined" (Edgeworth 1922, p. 267). This is the nub of the matter: in order for the posterior probability of any event to be positive, its prior probability must be positive (cf. Keynes 1921, p. 238). Within a year of Broad's 1918 paper, Wrinch and Jeffreys (1919) noted that the difficulty could be averted by using priors which place point masses at the endpoints of the unit interval.[17]

In 1919 this may have seemed to some to beg the question; after the ascendancy of the Ramsey/de Finetti approach to epistemic probability, it seems quite natural. Permitting passage from a unique uniform prior to a much wider class was crucial if this objection was to be successfully met.

What is the justification for assigning positive probability to the end points? The argument is in fact quite simple: *not* assuming some positive prior probability for the universal generalization is not an assumption of neutrality or absence of knowledge; it means that the universal generalization is assumed to have probability *zero*, i.e., we are *certain* it will not occur. Thus this classical objection is in fact a non-objection: unless one is *certain* that the universal generalization is false, its posterior probability increases with the number of confirming instances.

9. BRUNO DE FINETTI AND THE RIDDLE OF INDUCTION

Despite its virtues, the Wrinch–Jeffreys formulation suffers both from its continuing appeal to the "unknown chance" p, and the apparently ad hoc nature of the priors it suggests. Both of these defects were removed in 1931, when Bruno de Finetti proved his justly famous infinite representation theorem.[18] De Finetti showed that if an infinite sequence of 0–1 valued random variables $X_1, X_2, \ldots, X_n, \ldots$ is exchangeable for every n, then the limiting frequency of ones, $Z = \lim_{n \to \infty}(X_1 + X_2 + \cdots X_n)/n$, will exist with probability 1, and the probability distribution of the sequence can be represented as a mixture of binomial probabilities having this (random) limiting frequency Z as success parameter. Thus de Finetti provided a subjectivist account of objectivist chance and the role of parameters in statistics.

De Finetti also employed his infinite representation theorem, moreover, to provide a qualitative explanation for induction similar to and directly inspired by the French mathematician Henri Poincaré's *method of arbitrary functions* (Poincaré 1902, chap. 11). Poincaré had also sought to explain the existence of objective chance: to account for the apparently paradoxical fact that the outcome of tossing a coin, rolling a die, or spinning a roulette wheel results in equiprobable outcomes, despite our ignorance of and inability to control the complex physical conditions under which these occur. Poincaré was able to show that essentially independent of the distribution of physical inputs – in tossing a die, for example, the imparted linear velocity and angular momentum – the outcomes will occur with *approximately* equal probabilities for reasons that are purely mathematical. Detailed quantitative knowledge of input is thus unnecessary for approximate knowledge of outcome. Likewise, de Finetti was able to show for exchangeable sequences that, essentially independent of the initial distribution for the limiting frequency Z, after observing a sufficiently long initial segment of the sequence the posterior distribution of Z will always be highly peaked about the observed frequency p^*, and future trials expected to occur with a frequency very close to that of p^*.

De Finetti's is thus a *coherence* explanation of induction: if past and future are judged exchangeable, then – if we are to be consistent – we must expect future frequencies to resemble those of the past. But unlike the unique quantitative answer of Bayes, or the continuum of quantitative answers provided by Jeffreys, Wrinch, and Johnson, de Finetti's solution to Hume's problem of induction is a qualitative one. Whatever our prior beliefs, our inferences about the future based on our knowledge of the past will indeed be inductive in nature, but in ways that do not admit of unique numerical expression and may vary from person to person.

De Finetti's solution of Hume's problem of induction is a profound achievement, one of the few occasions when one of the deep problems of philosophy has yielded to direct attack. De Finetti's solution *can* be criticized, but such criticisms must go either to the nature of probable belief itself (can it be quantified? how does it change when new information is received?), or the illposed nature of Hume's problem (*how* is the future supposed to resemble the past?).

61

What is remarkable about de Finetti's essay *Probabilismo* is the clarity with which de Finetti saw these issues from the very beginning, and how closely they fit there into a unified view of science and philosophy. To many it may seem a strange and unfamiliar landscape; perhaps it "will only be understood by someone who has himself already had the thoughts that are expressed in it".[19]

Although the derivation of the finite rule of succession via exchangeability is both simple and elegant, the direct combinatorial proofs of the rule have a mathematical attractiveness of their own. We begin with the simplest case, considered by Broad, when all outcomes observed are successes.

Fix $m \geq 1$, and for $n \geq m$, let $a_n =: n!/(n - m)!$ and $S_n =: a_m + a_{m+1} + \cdots + a_n$.

Lemma 1. *For all $n \geq m$, $S_n = (n + 1)!/\{(m + 1)(n - m)!\}$.*

Proof. By induction. For $n = m$, $S_m = a_m = m! = (m + 1)!/\{(m + 1)(m - m)!\}$. Suppose next that the lemma is true for some $n \geq m$, and let

$$A_n = \frac{(n + 1)!}{(m + 1)(n - m)!}.$$

The induction hypothesis then states that $S_n = A_n$, and we wish to prove that $S_{n+1} = A_{n+1}$. But

$$
\begin{aligned}
S_{n+1} = S_n + a_{n+1} &= A_n + a_{n+1} \\
&= \frac{(n + 1)!}{(m + 1)(n - m)!} + \frac{(n + 1)!}{(n - m + 1)!} \\
&= (n + 1)! \frac{(n - m + 1) + (m + 1)}{(m + 1)(n - m + 1)!} \\
&= (n + 1)! \frac{(n + 2)}{(m + 1)(n - m + 1)!} \\
&= \frac{(n + 2)!}{(m + 1)(n - m + 1)!} = A_{n+1}.
\end{aligned}
$$

$\square$

The integer m was fixed in the above argument. If we now denote the dependence of S_n on m by $S_{n,m}$, then Broad's result follows by noting that the succession probability is the quotient $(n - m)^{-1} S_{n,m+1}/S_{n,m}$ and cancelling terms. Thus

$$P[X_{m+1} = 1 | X_1 = X_2 = \cdots = X_m = 1]$$

$$= \frac{\sum_{j=m+1}^{n} j(j-1)(j-2)\cdots(j-m)}{(n-m)\sum_{j=m}^{n} j(j-1)(j-2)\cdots(j-m+1)}$$

$$= \frac{1}{(n-m)} \frac{S_{n,m+1}}{S_{n,m}}$$

$$= \frac{1}{(n-m)} \frac{(n+1)!/\{(m+2)(n-m-1)!\}}{(n+1)!/\{(m+1)(n-m)!\}}$$

$$= \frac{(m+1)}{(m+2)}.$$

If instead of all m trials being successes, it is assumed that there are p successes and q failures, then the necessary sum becomes

$$\sum_j \frac{(n-q-j)!}{(n-m-j)!} \frac{(q+j)!}{j!},$$

and it may be again evaluated to be

$$\frac{(p!q!)}{(m+1)!} \frac{(n+1)!}{(n-m)!}.$$

Namely, if there are p successes and q failures, the possible urn compositions are

$$\mathbf{H}_j : (q+j) \text{ black and } (n-q-j) \text{ white}, 0 \leq j \leq n-m.$$

Under hypothesis $\mathbf{H}_j$ the probability that p whites and q blacks will be observed in the first m trials is

$$\mathbf{P}_j = \binom{m}{q} \frac{P_j Q_j}{(n)(n-1)\cdots(n-m+1)},$$

where

$$P_j = (n-q-j)(n-q-j-1)\cdots(n-q-j-p+1)$$

$$= \frac{(n-q-j)!}{(n-m-j)!}$$

and

$$Q_j = (q+j)(q+j-1)(q+j-2)\cdots(j+1) = \frac{(q+j)!}{(j)!}.$$

By Bayes's theorem, it follows that the posterior probability of $\mathbf{H}_i$, given that p whites and q blacks have been observed, is

$$P[\mathbf{H}_i|p, q] = \mathbf{P}_i(n + 1)^{-1} \Big/ \sum_j \mathbf{P}_j(n + 1)^{-1}$$

$$= \mathbf{P}_i \Big/ \sum_j \mathbf{P}_j = P_i Q_i \Big/ \sum_j P_j Q_j,$$

where

$$P_i Q_i = \frac{(n - q - i)!}{(n - m - i)!} \frac{(q + i)!}{i!},$$

and

$$\sum_j P_j Q_j = \sum_j \frac{(n - q - j)!}{(n - m - j)!} \frac{(q + j)!}{j!}.$$

Lemma 2.

$$\sum_{j=0}^{n-m} \frac{(n - q - j)!}{(n - m - j)!} \frac{(q + j)!}{j!} = p!q! \frac{(n + 1)!}{(m + 1)!(n - m)!}.$$

Proof. Dividing by $p!q!$ and denoting $n - m$ by k yields

$$\sum_{j=0}^{k} \binom{k + p - j}{k - j} \binom{q + j}{j} = \binom{k + p + q + 1}{k},$$

which is a well-known combinatorial identity (e.g., Whitworth 1897, p. xiii; Feller 1968, p. 65, problem 14 and p. 238, problem 15). $\square$

All of the classical derivations of the finite rule of succession reduce to the evaluation of this sum.

A.1. Prevost and L'Huilier's Proof

The Prevost–L'Huilier proof draws on the machinery of the "figurate numbers."

Let

$$f_0^j = 1, \quad j = 1, 2, 3, \ldots$$

and

$$f_i^j = \sum_{k=1}^{j} f_{i-1}^k, \quad i, j \geq 1;$$

the f_i^j are called the *figurate numbers* and are related to the binomial coefficients by the formula $f_i^j = \binom{j-i+1}{i}$; see generally Edwards (1987).

Prevost and L'Huilier prove the figurate identity

$$f_{p+q+1}^n = \sum_{i=1}^{n} f_q^i f_p^{n-i+1}$$

by induction on q, from which the corresponding result for binomial coefficients immediately follows. (Edwards 1987, p. 107, attributes this result in the special case $q = 1$ to Leibniz in 1673 but does not cite a source for the general result, his equation (8.24). Edwards's k, l, and s are our $p + q + 1$, n, and q respectively.)

Proof. If $q = 0$, then the identity is valid for all $p \geq 0$, since $f_0^i = 1$, all i, and the identity follows from the definition of the figurate numbers. Suppose the identity has been proved for some $q \geq 0$ and all $p \geq 0$. Then it holds for $q + 1$ and all $p \geq 0$ because

$$f_{p+q+2}^n = \sum_{j=1}^{n} f_q^j f_{p+1}^{n-j+1} \qquad \text{(induction step)}$$

$$= \sum_{j=1}^{n} f_q^j \left\{ \sum_{i=j}^{n} f_p^{n-i+1} \right\} \qquad \text{(definition of } f_i^j\text{)}$$

$$= \sum_{i=1}^{n} \left\{ \sum_{j=1}^{i} f_q^j \right\} f_p^{n-i+1} \qquad \text{(rearrangement)}$$

$$= \sum_{i=1}^{n} f_{q+1}^i f_p^{n-i+1} \qquad \text{(definition of } f_i^j\text{)};$$

the first step uses the induction hypothesis for q and $p + 1$.

Alternatively,

$$\sum_{i=1}^{n} f_{q+1}^i f_p^{n-i+1} = \sum_{i=1}^{n} \left\{ \sum_{j=1}^{i} f_q^j \right\} f_p^{n-i+1} \qquad \text{(definition of } f_i^j\text{)}$$

$$= \sum_{i=1}^{n} \left\{ \sum_{j=1}^{i} f_q^{i-j+1} \right\} f_p^{n-i+1} \qquad \text{(reverse summation)}$$

$$= \sum_{j=1}^{n} \sum_{i=j}^{n} f_q^{i-j+1} f_p^{n-i+1} \qquad \text{(rearrangement)}$$

65

$$= \sum_{j=1}^{n} \sum_{k=1}^{n-j+1} f_q^k f_p^{n-j-k+2} \qquad (k = i - j + 1)$$

$$= \sum_{j=1}^{n} f_{p+q+1}^{n-j+1} \qquad \text{(induction)}$$

$$= f_{p+q+2}^{n}. \qquad\qquad \square$$

This second proof is in fact the one given by Prevost and L'Huilier (although they only prove the specific cases $q = 0$, 1, 2, and 3, "la marche de la démonstration générale étant entièrement semblable à celle des exemples précédents, et ne présentant aucune difficulté" (p. 122)).

The fundamental binomial identity is also a special case of the more general result

$$\sum_{k=0}^{n} A_k(a, b) A_{n-k}(c, b) = A_n(a + c, b),$$

where $A_k(a, b) = (a/(a + bk))\binom{a+bk}{k}$, and a, b, c are arbitrary real numbers. As Gould and Kaucky (1966, p. 234) note, this identity "has been widely used, rediscovered repeatedly, and generalized extensively". Gould and Kaucky attribute it to H. A. Rothe, who proved it in his Leipzig dissertation of 1793, and thus appears to have priority over Prevost and L'Huilier.

A.2. Bishop Terrot's Proof

Bishop Terrot sums the series in Lemma 2 by using Abel partial summation (e.g., Knopp 1947, p. 313) and the identity

$$\sum_{j=0}^{a} \frac{(b + j)!}{j!} = \frac{(a + b + 1)(a + b) \cdots (a + 1)}{(b + 1)}.$$

If one denotes the sum in Lemma 2 by $A_{n,p,q}$, then it follows that

$$A_{n,p,q} = \left\{ \frac{q}{(p + 1)} \right\} A_{n,p+1,q-1},$$

and repeating the process a total of q times yields Lemma 2.

A.3. Todhunter's Proof

Todhunter remarks that the sum is readily obtained "by the aid of the binomial theorem" (Todhunter 1865, p. 454). By this he means comparing the coefficients in $(x + y)^{a+b} = (x + y)^a (x + y)^b$, with negative exponents permitted; see Feller (1968, p. 65, problem 14).

A.4. Ostrogradskii

The Russian mathematician M. V. Ostrogradskii also appears to have analyzed this problem in 1846; see Maistrov (1974, pp. 182–84).

A.5. Whipple's Proof

Jeffreys (1973, Appendix II) reports a simple counting argument proof. Combinatorial identities can often be profitably thought of as counting a certain quantity in two different ways, and Whipple discovered such an interpretation for Lemma 2.

NOTES

1. Laplace (1774; 1812, p. 402; 1814, p. xvii). The terminology is due to Venn, who gave this name to Chapter VIII of his *Logic of Chance*, adding in a footnote: "A word of apology may be offered here for the introduction of a new name. The only other alternative would have been to entitle the rule one of Induction. But such a title I cannot admit, for reasons which will be almost immediately explained" (Venn 1888, p. 190).
2. Broad later abandoned this assumption; see Broad (1927).
3. For further information about Prevost and L'Huilier, see the entries on both in the *Dictionary of Scientific Biography*.
4. The reader should be cautioned the literature abounds with confusions between the finite and infinite cases, and the cases involving sampling with and without replacement. For example, Keynes (1921, p. 378) states that "the rule of succession does not apply, as it is easy to demonstrate, even to the case of balls drawn from an urn, if the number of balls is finite" (Keynes 1921, p. 378). Likewise Strong (1976, p. 203) confuses the answers for sampling with and without replacement from a finite urn, a confusion that may stem in part from an unfortunate typographical error in Boole.
5. Stigler's nomination of Nicholas Saunderson, Lucasian Professor of Mathematics at Cambridge from 1711 to 1739, as a plausible alternative candidate (Stigler 1983), cannot be seriously credited. There is no evidence that Saunderson ever wrote on any topic in probability. Hartley's wording, moreover ("an ingenious friend has communicated to me . . . "), suggests information recently received from a living person, rather than a friend long dead (Saunderson had died a decade earlier, in 1739). The anonymity employed would scarcely make sense otherwise. Stigler's concluding "Bayesian calculation" (summarized in his Table 1), purporting to show that the odds are 3 to 1 in favor of Saunderson over Bayes, curiously omits the single most important item of evidence in favor of Bayes – that Bayes is known to have written a manuscript on the subject, while Saunderson is not.

6. In his original memoir on inverse probability, where the rule of succession is first stated, Laplace begins by considering an urn "supposed to contain a given number of white and black tickets in an unknown ratio", but in his solution he assumes that the urn contains "an infinity of white and black tickets in an unknown ratio". Strictly construed, of course, this latter statement makes no sense, and is clearly intended as an abbreviated way of saying something else. De Morgan considers that the contents of the urn are assumed infinite "only that the withdrawal of a definite number may not alter the ratio" (1847, p. 214), and he goes on to note that if the contents of the urn are finite, but the sampling is *with* replacement, then as the number of balls increases, the resulting rule of succession will approximate the Laplacean answer of $(m + 1)/(m + 2)$. Bishop Terrot, on the other hand, uses the expression "infinite or indefinite" in describing this case (Terrot 1853, p. 543), and clearly takes the reference to infinite contents to be a circumlocution for the asymptotic, large-sample result. The paper by Prevost and L'Huilier was an attempt to clarify matters by determining the exact, small-sample answer, and it must have come as a considerable surprise to find that there was no difference between the two. The philosophical importance of the result is that, whatever its other real or alleged defects, Laplace's analysis cannot be faulted on the grounds of its appeal to the infinite. This point is sometimes overlooked.

7. "... it was Hume who furnished the Laplacean school with its philosophy of science," as Stove (1973, p. 102) notes. Hume's influence, especially on Condorcet, is ably discussed by Baker (1975, chap. 3, passim), and appears most clearly in the work of Condorcet's disciple, Sylvestre-Francois Lacroix (Stove 1973, p. 103; Baker 1975, pp. 186–87).

8. Strictly speaking, Price misstates the rule in two ways: (a) the odds are $2^n - 1$ to 1, not 2^n to 1; (b) the exponent of 2 should be the number of risings, not the number of returns. (Thus the true odds are $2^{1,000,001} - 1$ to 1.) Price gives the correct formula, however, earlier in his appendix.

9. The French naturalist Buffon gives the 2^n to 1 rule in his *Essai d'arithmétique morale* of 1777; taking the age of the earth to be 6,000 years, he concludes the probability that the sun will rise the following day is 22,189,999 to 1. Although Buffon does not cite the rule's source, it is clearly taken from Bayes's essay: Price's fanciful narrative of a man who observes the rising of the sun for the first time has been lifted, without attribution, virtually word-for-word! (Zabell 1988).

10. Likewise, the English mathematician Waring (1794, p. 35) dissented from the identification:

I know that some mathematicians of the first class have endeavoured to demonstrate the degree of probability of an event's happening n times from its having happened m preceding times; and consequently that such an event will probably take place; but, alas, the problem far exceeds the extent of human understanding; who can determine the time when the sun will probably cease to run its present course?

11. The Truscott and Emory translation of the *Essai* renders "principe régulateur" as "principal regulator". This is not only incorrect, it introduces a deistic note entirely foreign to Laplace, and obscures the essential point of the passage.

12. The level of English mathematics was at ebbtide at the beginning of the 19th century, and De Morgan was one of a group of English mathematicians and scientists (including Babbage, Herschel, and Galloway) who attempted to remedy the situation during the first half of the century through a series of popular and technical tracts.

13. The mathematical machinery for the generalization is provided by Laplace (*Théorie*, p. 376), although it is not stressed. The uniform prior on the unit interval is replaced by the uniform prior on the simplex $\Delta_t =: \{(p_1, p_2, \ldots, p_t) : p_1, p_2, \ldots, p_t \geq 0, p_1 + p_2 + \cdots + p_t = 1\}$, and the rule of succession becomes $(m_i + 1)/(M + t)$, where m_i is the number of outcomes in the ith category. (This is, of course, nothing other than Carnap's c^*.) De Morgan discusses the rule of succession in this more general context (De Morgan 1838, pp. 66–69; 1845, pp. 413–14), including the so-called "sampling of species problem", where one does not know the total number of categories.

14. Now the inappropriateness of the application of the rule of succession to the rising of the sun becomes manifest: successive risings are not exchangeable. (For example, although for most people the probability that the sun will rise tomorrow, but not rise the day after is quite small, the probability that it will not rise tomorrow but will rise the day after is much smaller still.)

15. See generally Zabell (1982).

16. For discussion of Fries's analysis, see Krüger (1987, pp. 67–68).

17. The objection that simply because the limiting frequency of 1s in the sequence can equal 1, it does not follow that all elements of the sequence will equal 1, is easily met by working at the level of sequence space and giving the sequences $\{1, 1, 1, \ldots\}$ and $\{0, 0, 0, \ldots\}$ positive probability.

18. The following briefly summarizes an argument given in much greater detail in Zabell (1988).

19. I thank Persi Diaconis for a number of helpful discussions over the years regarding finite exchangeability (in particular, for pointing out the identity of the Prevost–L'Huilier and the Bayes–Laplace processes).

REFERENCES

Baker, Keith Michael 1975. *Condorcet: From Natural Philosophy to Social Mathematics*, Chicago: University of Chicago Press.

Bayes, Thomas 1764. 'An Essay Towards Solving a Problem in the Doctrine of Chances', *Philosophical Transactions of the Royal Society of London* **53**, 370–418, reprinted in E. S. Pearson and M. G. Kendall (eds.), *Studies in the History of Statistics and Probability*, Vol. 1, London: Charles Griffin, 1970, pp. 134–53, page references are to this edition.

Bernoulli, Jakob 1713. *Ars conjectandi*, Thurnisiorum, Basel, reprinted in *Die Werke von Jakob Bernoulli*, Vol. 3, Basel: Birkhauser, 1975, pp. 107–286.

Bertrand, J. 1907. Calcul des probabilités, Paris: Gauthier-Villars, (1st ed., 1889).

Bolzano, Bernard 1837. Wissenschaftslehre, translated 1972 under the title *Theory of Science*, R. George (ed. and trans.), Berkeley: University of California Press.

Boole, George 1854. *An Investigation of the Laws of Thought*, London: Macmillan, reprinted 1958, New York: Dover.

Broad, C. D. 1918. 'The Relation Between Induction and Probability', *Mind* **27**, 389–404; **29**, 11–45.

Broad, C. D. 1922. 'Critical Notice on J. M. Keynes', *A Treatise on Probability, Mind* **31**, 72–85.

Broad, C. D. 1924. 'Mr. Johnson on the Logical Foundations of Science', *Mind* **33**, 242–61, and 369–84.

Broad, C. D. 1927. 'The Principles of Problematic Induction', *Proceedings of the Aristotelian Society* **28**, 1–46.

Carnap, Rudolph 1950. *Logical Foundations of Probability*, 2nd ed. 1962, Chicago: University of Chicago Press.

Carnap, Rudolph 1952. *The Continuum of Inductive Methods*, Chicago: University of Chicago Press.

Condorcet, Le Marquis de 1785. *Essai sur l'application de l'analyse à la probabilité dés décisions rendues à la pluralité des voix*, Paris: Imprimerie Royale.

Cournot, Antoine Augustin 1843. *Exposition de la théorie des chances et des probabilités*, Libraire de L. Hachette, Paris.

Dale, A. I. 1986. 'A Newly-Discovered Result of Thomas Bayes', *Archive for History of Exact Sciences* **35**, 101–13.

De Finetti, Bruno 1937. 'La prevision: ses lois logiques, ses sources subjectives', *Annales de l'Institut Henri Poincaré* **7**, 1–68, translated in H. E. Kyburg, Jr. and H. E. Smokler, (eds.), *Studies in Subjective Probability*, New York: Wiley, 1964, pp. 93–158, page references are to this edition.

De Morgan, Augustus 1838. *An Essay on Probabilities, and their Application to Life Contingencies and Insurance Offices*, London: Longman, Orme, Brown, Green, and Longmans.

De Morgan, Augustus 1845. 'Theory of Probabilities', *Encyclopedia Metropolitana, Vol. 2: Pure Mathematics*, pp. 393–490, London: B. Fellowes et al.

De Morgan, Augustus 1847. *Formal Logic: Or the Calculus of Inference Necessary and Probable*, London: Taylor and Watton.

Diaconis, Persi 1977. 'Finite Forms of de Finetti's Theorem on Exchangeability', *Synthese* **36**, 271–81.

Edgeworth, Francis Ysidro 1884. '*A Priori* Probabilities', *Philosophical Magazine* (Series 5) **18**, 205–10.

Edgeworth, Francis Ysidro 1922. 'The Philosophy of Chance', *Mind* **31**, 257–83.

Edwards, A. W. F. 1978. 'Commentary on the Arguments of Thomas Bayes', *Scandinavian Journal of Statistics* **5**, 116–18.

Edwards, A. W. F. 1987. *Pascal's Arithmetic Triangle*, New York: Oxford University Press.

Feller, William 1968. *An Introduction to Probability Theory and its Applications*, Vol. 1, 3rd ed., New York: Wiley.

Fisher, Ronald A. 1973. *Statistical Methods and Scientific Inference*, 3rd ed., (1st ed., 1956; 2nd ed., 1959), New York: Hafner Press.

Good, Irving John 1950. *Probability and the Weighing of Evidence*, New York: Hafner Press.

Good, Irving John 1965. *The Estimation of Probabilities: An Essay on Modern Bayesian Methods*, Cambridge, MA: MIT Press.

Goodman, Nelson 1979. *Fact, Fiction, and Forecast*, 3rd ed., Indianapolis: Hackett.

70

Gould, H. W. and J. Kaucky 1966. 'Evaluation of a Class of Binomial Coefficient Summations', *Journal of Combinatorial Theory* **1**, 233–47.

Hacking, Ian 1975. *The Emergence of Probability*, New York: Cambridge University Press.

Hartley, David 1749. *Observations on Man, his Frame, his Duty, and his Expectations*, London: S. Richardson.

Heath, P. L. 1967. 'Jevons, William Stanley', in P. Edwards (ed.), *The Encyclopedia of Philosophy*, Vol. 4, New York: Macmillan, pp. 260–61.

Hume, David 1739. *A Treatise of Human Nature*, London. Page references are to the 2nd edition of the L. A. Selbe-Bigge text, revised by P. H. Nidditch, Oxford, UK: Clarendon Press, 1978.

Jeffreys, Harold 1939. *The Theory of Probability*, Oxford, UK: Clarendon Press, (2nd ed. 1958; 3rd ed. 1961).

Jeffreys, Harold 1973. *Scientific Inference*, 3rd ed., New York: Cambridge University Press.

Jevons, William Stanley 1874. *The Principles of Science: A Treatise on Logic and Scientific Method*, 2 vols., London: Macmillan, (2nd ed., 1877), reprinted 1958, New York: Dover.

Johnson, William Ernest 1924. *Logic, Part III: The Logical Foundations of Science*, Cambridge: Cambridge University Press.

Johnson, William Ernest 1932. 'Probability: The Deductive and Inductive Problems', *Mind* **41**, 409–23.

Kamlah, Andreas 1983. 'Probability as a Quasi-Theoretical Concept: J. v. Kries' Sophisticated Account After a Century', *Erkenntnis* **19**, 239–51.

Kamlah, Andreas 1987. 'The Decline of the Laplacian Theory of Probability: A Study of Stumpf, von Kries, and Meinong', in L. Kruger, L. J. Daston, and M. Heidelberger (eds.), *The Probabilistic Revolution, Volume 1: Ideas in History*, Cambridge, Mass: MIT Press, pp. 91–110.

Keynes, John Maynard 1921. *A Treatise on Probability*, London: Macmillan.

Kneale, William 1949. *Probability and Induction*, Oxford, UK: The Clarendon Press.

Knopp, Konrad 1947. *Theory and Application of Infinite Series*, New York: Hafner Press.

Kries, Johann von 1886. *Die Principien der Wahrscheinlichkeitsrechnung*, Tübingen: Mohr, (2nd ed., 1927).

Krüger, Lorenz 1987. 'The Slow Rise of Probabilism', in L. Krüger, L. J. Daston, and M. Heidelberger (eds.), *The Probabilistic Revolution, Volume 1: Ideas in History*, Cambridge, Mass: MIT Press, pp. 59–89.

Laplace, Pierre Simon Marquis de 1774. 'Mémoire sur la probabilité des causes par les évenements', *Mémoires de l'Académie royale des sciences présentés par divers savans* **6**, 621–56, reprinted in *Oeuvres complètes de Laplace* (1878–1912), Vol. 8, pp. 27–65, Paris: Gauthier-Villars, translated in Stigler (1986).

Laplace, Pierre Simon Marquis de 1812. *Théorie analytique des probabilités*, Courcier, Paris, 2nd ed., 1814; 3rd ed., 1820, page references are to *Oeuvres complètes de Laplace*, Vol. 7, Paris: Gauthier-Villars, 1886.

Laplace, Pierre Simon Marquis de 1814. *Essai philosophique sur les probabilités*, Courcier, Paris, page references are to *Oeuvres complètes de Laplace*, vol. 7, Paris: Gauthier-Villars, 1886.

Laudan, L. 1973. 'Induction and Probability in the Nineteenth Century', in P. Suppes et al. (eds.), *Logic, Methodology and Philosophy of Science IV*, Amsterdam: North-Holland, pp. 429–38.

Madden, Edward H. 1960. 'W. S. Jevons on Induction and Probability', in E. H. Madden (ed.), *Theories of Scientific Method: The Renaissance Through the Nineteenth Century*, Seattle: University of Washington Press, pp. 233–47.

Maistrov, L. E. 1974. *Probability Theory: A Historical Sketch*, New York: Academic Press.

Mill, John Stuart 1843. *A System of Logic*, 2 Vols., London: John W. Parker.

Ostrogradskii, M. V. 1848. 'On a Problem Concerning Probabilities', *St. Petersburg Academy of Sciences* **6**, 321–46, reprinted in M. V. Ostrogradskii, *Polnoe sobranie trudov*, Vol. 3, Academy of Sciences URRSSR, Kiev, 1961.

Pearson, Karl 1978. *The History of Statistics in the 17th and 18th Centuries*, E. S. Pearson (ed.), New York: Macmillan.

Poincaré, Henri 1902. *La science et l'hypothèse,* Paris, translated 1905 as *Science and Hypothesis*, J. W. Greenstreet (trans.), Walter Scott, London; translation reprinted 1952, New York: Dover.

Prevost, Pierre and S. A. L'Huilier 1799. 'Sur les probabilités', *Mémoires de l'Academie Royale de Berlin* 1796, pp. 117–42.

Prevost, Pierre and S. A. L'Huilier 1799a. 'Mémoire sur l'art d'estimer la probabilité des causes par les effets', *Mémoires de l'Academie Royale de Berlin* 1796, pp. 3–24.

Price, Richard 1758. *A Review of the Principal Questions in Morals*, London (2nd ed., 1769; 3rd ed., 1787), reprinted 1974, D. D. Raphael (ed.), Oxford, UK: Clarendon Press, page references are to the 1974 edition.

Ramsey, Frank Plumpton 1926. 'Truth and Probability', in R. B. Braithwaite (ed.), *The Foundations of Mathematics and other Logical Essays*, London: Routledge and Kegan Paul, 1931, pp. 156–98.

Stigler, Stephen M. 1982. 'Thomas Bayes's Bayesian Inference', *Journal of the Royal Statistical Society Series A* **145**, 250–58.

Stigler, Stephen M. 1983. 'Who Discovered Bayes's Theorem?', *American Statistician* **37**, 290–96.

Stigler, Stephen M. 1986. 'Laplace's 1774 Memoir on Inverse Probability', *Statistical Science* **1**, 359–78.

Stove, D. C. 1973. *Probability and Hume's Inductive Scepticism*, Oxford, UK: Clarendon Press.

Strong, John V. 1976. 'The Infinite Ballot Box of Nature: De Morgan, Boole, and Jevons on Probability and the Logic of Induction', *PSA 1976: Proceedings of the Philosophy of Science Association* **1**, 197–211.

Strong, John V. 1978. 'John Stuart Mill, John Herschel, and the 'Probability of Causes', *PSA 1978: Proceedings of the Philosophy of Science Association* **1**, 31–41.

Terrot, Bishop Charles 1853. 'Summation of a Compound Series, and its Application to a Problem in Probabilities', *Transactions of the Edinburgh Philosophical Society* **20**, 541–45.

Todhunter, Isaac 1865. *A History of the Mathematical Theory of Probability from the Time of Pascal to that of Laplace*, London: Macmillan, reprinted 1965, New York: Chelsea.

Venn, John 1866. *The Logic of Chance*, London: Macmillan (2nd ed., 1876; 3rd ed., 1888), reprinted 1962, New York: Chelsea.

von Wright, Georg Henrik 1957. *The Logical Problem of Induction*, 2nd revised edition, New York: Macmillan.

Waring, E. 1794. *An Essay on the Principles of Human Knowledge*, Cambridge: Deighton Bell.

Whitworth, William Allen 1901. *Choice and Chance*, 5th ed., Cambridge: Deighton Bell.

Wrinch, Dorothy and Harold Jeffreys 1919. 'On Certain Aspects of the Theory of Probability', *Philosophical Magazine* **38**, 715–31.

Zabell, Sandy L. 1982. 'W. E. Johnson's "Sufficientness" Postulate', *Annals of Statistics* **10**, 1091–99.

Zabell, Sandy L. 1988. 'Symmetry and its Discontents', in Harper, W. and Skyrms, B. (eds.), *Probability, Chance, and Causation*, Vol. I, Dordrecht: Kluwer, pp. 155–190.

Zabell, Sandy L. 1988a. 'Buffon, Price, and Laplace: Scientific Attribution in the 18th Century', *Archive for History of Exact Sciences* **39**, 173–81.

Zabell, Sandy L. 1989. 'R. A. Fisher on the History of Inverse Probability', *Statistical Science*, **4,** 247–63.

3

Buffon, Price, and Laplace: Scientific Attribution in the 18th Century

1. INTRODUCTION

Laplace's rule of succession states, in brief, that the probability of an event recurring, given that it has already occurred n times in succession, is $(n + 1)/(n + 2)$.[1] In his *Essai philosophique sur les probabilités* (1814), Laplace gave a famous, if notorious illustration of the rule: the probability of the sun's rising.

Thus we find that an event having occurred successively any number of times, the probability that it will happen again the next time is equal to this number increased by unity divided by the same number, increased by two units. Placing the most ancient epoch of history at five thousand years ago, or at 1826213 days, and the sun having risen constantly in the interval at each revolution of twenty-four hours, it is a bet of 1826214 to one that it will rise again to-morrow. [Laplace, *Essai philosophique*, p. xvii]

This passage was at the center of a spirited debate for over a century about the ability of the calculus of probabilities to provide a satisfactory account of inductive inference (e.g., Keynes 1921, Chapter 30). Although the later history of this debate is well known, what is less well known, perhaps, is its history prior to the appearance of Laplace's *Essai*. In fact, the question whether belief in the future rising of the sun can be expressed probabilistically had been briefly alluded to by Hume in his *Treatise* of 1739, and had been discussed prior to the appearance of Laplace's *Essai* by Price, Buffon, Condorcet, Waring, Prevost, and L'Huilier (e.g., Zabell, 1988, Section 5).[2]

The only hint that Laplace gives of this prior debate is a brief, cryptic reference to a formula differing from the rule of succession, which had been propounded by the famous French naturalist Buffon in 1777. Laplace notes Buffon's alternative solution, only to dismiss it with the comment, "the true manner of relating past events with the probability of causes and of future events was unknown to this illustrious writer."

Reprinted with permission from *The Archive for History of Exact Science 39* 2 (1988): 173–181.

Buffon gave no indication of where his answer came from – it is simply stated without proof – and subsequent commentators have shared Laplace's (apparent) opinion that it is simply erroneous, a relic of the early days of mathematical probability, when confusion and error abounded. Todhunter, for example, in his *History of the Mathematical Theory of Probability*, says that Buffon "lays down without explanation a peculiar principle" which is "quite arbitrary", and cites the passage from Laplace quoted above (Todhunter, 1865, p. 344). Karl Pearson likewise, in his lectures on the history of statistics, notes that Buffon "does not explain on what hypothesis he bases his" answer, and that "it is clearly not consistent with Laplace's" (Pearson, 1978, pp. 193–194). Curiously, as will be seen below, both Todhunter and Pearson elsewhere in their books discussed the source of Buffon's answer, but failed to make the connection. Tracing the roots of Buffon's answer is an interesting historical detective story, which sheds some light on attribution practices in the eighteenth century.

2. BUFFON'S *ESSAI D'ARITHMÉTIQUE MORALE*

The passage that Laplace alludes to is at the beginning of Buffon's *Essai d'arithmétique morale* of 1777. Buffon asks us to suppose a hypothetical person who has never before seen or heard anything, and to consider how belief or doubt will arise in him. If such a person were to see the sun rise and set for the first time, Buffon argues, he could conclude nothing beyond what he had already seen.[3] But if he saw the sun rise and set a second time, this would be a first "experience" [*cette second vision est une première expérience*] which would produce in him the expectation of seeing the sun again, an expectation which would be strengthened if he were to see the sun rise and set yet a third and fourth time; so that when he had seen the sun appear and disappear regularly many times in succession, he would be certain that it would continue to do so.[4] The greater the number of similar observations, moreover, the greater the certainty of seeing the sun rise again; each observation producing a probability, the sum of which eventually results in physical certitude.[5] For example, in 6,000 years, the sun will have risen 2,190,000 times and reckoning from the second day it rose, the probabilities of it rising the next day will augment like the sequence 1, 2, 4, 8, 16, 32, 64, ... or 2^{n-1}, where n is the number of days the sun has risen.[6]

Although it is not immediately apparent what Buffon means by a probability that augments as the sequence 1, 2, 4, ..., 2^{n-1}, Pearson (1978, p. 193) interprets this to mean that the odds in favor of the event are 2^{n-1} to 1, and discussion elsewhere in Buffon's essay confirms that this is indeed Buffon's meaning.[7]

Where does Buffon's answer come from? It clearly does not agree with Laplace's rule of succession, first published three years before (Laplace, 1774). The only earlier paper dealing with the problem of inverse probability is that of Bayes (1764), and it turns out that this is in fact the source of Buffon's answer, albeit somewhat garbled in the process of transmission.

3. BAYES'S *ESSAY TOWARDS SOLVING A PROBLEM IN THE DOCTRINE OF CHANCES*

In his "Essay towards solving a problem in the doctrine of chances" (1764), Bayes considers "an event concerning the probability of which we absolutely know nothing antecedently to any trials made concerning it." Employing an argument often unappreciated for its subtlety,[8] Bayes concluded that if the unknown probability of such an event is x, then the prior probability of x may be taken to be uniform on the unit interval. Bayes's essay was published posthumously by his friend the Reverend Dr. Richard Price, who added a cover letter explaining the purpose of the essay and an appendix "containing an application of the foregoing rules to some particular cases." One of these was to note that "if an event has happened n times, there will be an odds of $2^{n+1} - 1$ to one, for more than an equal chance that it will on further trials."[9]

This is *not* Laplace's rule of succession, but rather a calculation of the posterior probability that the unknown chance x of the event exceeds $\frac{1}{2}$, based on Bayes's assumption that all values of x are *a priori* equally likely.[10] Thus

$$P\left[x > \frac{1}{2}\right] = \int_{\frac{1}{2}}^{1} x^n dx \bigg/ \int_{0}^{1} x^n dx = 1 - (1/2)^{n+1} = (2^{n+1} - 1)/2^{n+1}.$$

If, like Buffon, one uses $n - 1$ in the exponent instead of n (since he considers the first trial merely to inform us that the event is possible), and one rounds the resulting odds of $2^n - 1$ to 1 up to 2^n to 1, this gives us a rule of 2^n to 1, essentially Buffon's rule. (Buffon's answer is actually 2^{n-1} to 1; the reason for the discrepancy will become apparent shortly.)

Buffon, it would seem, is quoting the Bayes/Price result. Is it possible that Buffon has independently derived the result? The answer turns out to be no. Price also discusses the rising of the sun, and all becomes clear when we contrast Buffon's treatment with his:

Let us imagine to ourselves the case of a person just brought forth into this world, and left to collect from his observation of the order and course of events what powers and causes take place in it. The Sun would, probably, be the first object that would engage his attention; but after losing it the first night he would be entirely ignorant whether he should ever see it again. He would therefore be in the condition of a person making a first experiment about an event entirely unknown to him. But let him see a second appearance or one return of the Sun, and an expectation would be raised in him of a second return, and he might know that there was an odds of 3 to 1 for some probability of this. These odds would increase, as pointed out before, with the number of returns to which he was witness. But no finite number of returns would be sufficient to produce absolute or physical certainty. For let it be supposed that he has seen it return at regular and stated intervals a million times. The conclusions this would warrant would be such as follow. There would be the odds of the millioneth power of 2, to one, that it was likely that it would return again at the end of the usual interval.

Clearly Buffon has taken his discussion – almost word-for-word – from that of Price![11] The former is merely a paraphrase of the latter. The quaint narrative of a person who had never seen the sun before, the setting aside of the first observation, the reference to physical certitude – all these have been copied, without attribution.[12] (Buffon adopts a different stand, however, on the issue of physical certitude.)

The discrepancies in exponent, moreover, now become clear. Price had earlier (and correctly) stated the odds in question to be $2^{n+1} - 1$ to one, where n is the number of times the event in question has *occurred*. In his discussion of the rising of the sun, however, Price implicitly (and incorrectly) gives the odds as 2^n to one. True, if n were the number of *risings*, then by Price's logic the exponent in the formula would indeed be n, since the number of occurrences $+ 1$ (the correct exponent) $=$ number of returns $+ 1 =$ numbers of risings. Thus, if the number of *risings* were a million, the (rounded) odds are $2^{1,000,000}$ to one. But since Price says *returns*, his answer is incorrect (it should be $2^{1,000,001}$ to one). Because of this error, superficial reading of the passage might well leave one with the mistaken impression that the appropriate exponent is the number of returns; that is, the number of occurrences.

This, then, is the source of Buffon's confusion: on the basis of this passage from Price, Buffon believes the exponent in the formula to be the number of occurrences, and in turn subtracts 1 from *his* n, the number of risings, to obtain incorrectly $2^{risings-1}$: 1 as the appropriate odds.[13]

Attribution and citation of the works of one's predecessors could be an uncertain affair in the eighteenth century. Laplace himself has been harshly criticized on occasion for failure to cite such work. Augustus De Morgan, for example, one of Laplace's most enthusiastic followers in England, wrote of Laplace's solution of the gambler's ruin problem:

The solution of Laplace gives results for the most part in precisely the same form as those of De Moivre, but, according to Laplace's usual custom, no predecessor is mentioned. Though generally aware that Laplace (and too many others, particularly among French writers) was much given to this unworthy species of suppression, I had not any idea of the extent to which it was carried out until I compared his solution of the problem of the duration of play, with that of De Moivre. Having been instrumental (in my mathematical treatise on Probabilities, in the *Encyclopedia Metropolitana*) in attributing to Laplace more than his due, having been mislead by the suppressions aforesaid, I feel bound to take this opportunity of requesting any reader of that article to consider every thing there given to Laplace as meaning simply that it is to be found in his work, in which, as in the *Mécanique Céleste*, there is enough originating from himself to make any reader wonder that one who could so well afford to state what he had taken from others, should have set an example so dangerous to his own claims. [De Morgan, 1838, Appendix, pp. i–ii.][14]

One could equally wonder about Buffon; the *Essai d'arithmétique morale* is a work of considerable originality and interest, and the passage we have discussed appears a curiosity.

Laplace's failure to cite his predecessors, however, is a complex affair, often bordering on questions of style.[15] The passage referring to Buffon's analysis of the rising of the sun furnishes an instance in point. We have not yet quoted Laplace's actual statement of Buffon's rule. What Laplace says is: "Buffon in his *Political Arithmetic* calculates differently the preceding probability. He supposes that it differs from unity only by a fraction whose numerator is unity and whose denominator is the number 2 raised to a power equal to the number of days which have elapsed since the epoch."

It would thus appear that Laplace almost certainly had Price's discussion in mind when he referred to Buffon. He states Buffon's rule to be $1 - (1/2)^n$, where n is the number of risings.[16] This is neither Buffon's rule (which was a garbled version of Price's), nor Price's rule (which was a garbled version of Bayes's), but the correct formula based on Bayes's postulate setting aside the first observation! Laplace has silently corrected both sets of errors in the formula. His point is that even the correct formula does not give the desired probability.[17]

Many of the examples in Laplace's *Théorie analytique* and *Essai philosophique* were similarly designed to correct misunderstandings and errors in previous work. It is only seldom, however, that Laplace directly refers to this literature. A good example in point is Laplace's discussion of testimony in Chapter 11 of the *Essai philosophique*.

Laplace's initial discussion centers around two simple cases: 1) a number is drawn from an urn containing the numbers 1–1,000; 2) a ball is drawn from an urn containing 999 black and 1 white balls. In each case it is assumed that a witness correctly announces the result 9 out of 10 times. In the first case Laplace calculates that the probability that the number 79 was drawn, given that it is stated to have been by the witness, is close to one; in the second, Laplace calculates that the probability that the ball drawn is white, given that it is stated to have been by the witness, is close to zero.

Laplace's analysis was initially faulted by both Mill and Venn, each of whom in later editions of their books grudgingly conceded that Laplace's analysis is correct in the circumstances he posits. Mill, for example, agrees it is "irrefragable in the case which he supposes, and in all others which that case fairly represents" (*Logic*, Book 3, Chapter 25, Section 6), but complains that it is not "a perfect representative of all cases of coincidence."[18]

But this was a claim Laplace had never advanced. Instead, as in his discussion of the rising of the sun, his examples had the much narrower purpose of clarifying issues that had previously arisen in a highly confused debate about testimonial reliability. Once again, Hume was a focal point of the debate. In his famous essay *On Miracles*, Hume had argued that in assessing testimony, one should weigh both the prior improbability of the fact attested and the reliability of the witness testifying. From a modern Bayesian viewpoint nothing could be more natural, but during the 18th century this claim was a point of considerable controversy. The Scottish divine George Campbell, for example, noted that we often hear reports of improbable events whose veracity we ordinarily never doubt (e.g., a witness tells us that a ferry which has successfully crossed a channel 1,000 times before has just sunk). Price, the editor of Bayes's manuscripts, noted as a particularly simple instance of this phenomenon the drawing of a specific number in a lottery (i.e., Laplace's example (1)). Condorcet attempted the first mathematical analysis of the problem, but used a formula only appropriate to Laplace's example (2), and his attempt to resolve the difficulties that arose only led to further confusion.[19]

Laplace's examples were clearly chosen with this past history in mind, and his discussion is the first careful and correct analysis of the differences between the two cases: why one should remain sceptical of the reports of certain types of improbable events (such as miracles), but not doubt the report of other, apparently equally improbable events (such as lottery drawings). But about the debate itself Laplace is entirely silent. He is content to give the correct analysis, and draw the proper conclusions, but he disdains to catalogue the history of error.[20]

NOTES

1. Laplace (1774); see also Laplace (1812, p. 402), cited below as *Théorie analytique*, and Laplace (1814, p. xvii), cited below as *Essai philosophique*. The terminology is due to Venn (1888, p. 190).

2. Hume in his *Treatise* of 1739 asserted: "One wou'd appear ridiculous, who wou'd say, that 'tis only probable the sun will rise-to morrow, or that all men myst dye; tho' 'tis plain we have no further assurance of these facts, than what experience affords us" (Hume, 1739, p. 124).

3. *En supposant un homme qui n'eût jamais rien vu, rien entendu, cherchons comment la croyance & le doute se produiroient dans son esprit; supposons-le frappé pour la première fois par l'aspect du soleil; il le voit briller au haut des Cieux, ensuite décliner & enfin disparoître; qu'en peut-il conclure? rien, sinon qu'il a vu le soleil, qu'il l'a vu suivre une certaine route, & qu'il ne le voit plus; . . .*

4. *. . . mais cet astre reparoît & disparoît encore le lendemain; cette seconde vision est une première expérience, qui doit produire en lui l'espérance de revoir le soleil, & il commence à croire qu'il pourroit revenir, cependant il en doute beaucoup; le soleil reparoît de nouveau; cette troisième vision fait une seconde expérience qui diminue le doute autant qu'elle augmente la probabilité d'un troisième retour; une troisième expérience l'augmente au point qu'il ne doute plus guère que le soleil ne revienne une quatriéme fois; & enfin quand il aura vu cet astre de lumière paroître & disparoître régulièrement dix, vingt, cent fois de suite, il croira être certain qu'il le verra toujours paroître, disparoître & se mouvoir de la même façon; . . .*

5. *. . . plus il aura d'observations semblables, plus la certitude de voir le soleil se lever le lendemain sera grande; chaque observation, c'est-à-dire, chaque jour, produit une probabilité, & la somme de ces probabilités réunies, dès qu'elle est trés-grande, donne la certitude physique; l'on pourra donc toujours exprimer cette certitude par les nombres, en datant de l'orgine du temps de notre expérience, & il en sera de même de tous les autres effets de la Nature; . . .*

6. *. . . par exemple, si l'on veut réduire ici l'ancienneté du monde & de notre expérience à six mille ans, le soleil n'est levé pour nous [footnote omitted] que 2 millions 190 mille fois, & comme à dater du second jour qu'il s'est levé, les probabilités de se lever le lendemain augmentent, comme la suite* $1, 2, 4, 8, 16, 32, 64, \ldots$ *ou* 2^{n-1}. *On aura (lorsque dans la suite naturelle des nombres, n est égale* 2,190000), *on aura, dis-je,* $2^{n-1} = 2^{2,189999}$; . . . [Buffon, 1777, p. 458].

7. In Section 8 of the *Essai*, Buffon says " . . . $2^{13} = 8192, \ldots$ & par conséquent lorsque cet effet est arrivé treize fois, il y a 8192 à parier contre 1, qu'il arrivera une quatorième fois . . . "

8. See, *e.g.*, Murray (1930), Edwards (1978), Stigler (1982), Stigler (1986).

9. Price denotes the number of trials by p; for consistency of notation this has been changed to an n in the quotation.

10. The reader is cautioned that there is great confusion in the literature about Price's calculation, and its relation to the rule of succession; see, *e.g.*, Todhunter (1865, p. 299), Pearson (1978, pp. 366–369), Dale (1982, pp. 43–45). Price's language, however, is quite clear.

    The great 19th century British scientist J. W. F. Herschel was one of the few to understand the distinction involved; see Herschel (1857, pp. 414–415).

11. Buffon would thus appear to be an exception to Stigler's statement that "Bayes' memoir passed unnoticed on the Continent" (Stigler,1978, p. 245). Gouraud (1848, p. 54) reports without indicating his source that Buffon's essay was composed approximately in 1760, and Buffon quotes in a footnote to Section 8 of his essay a letter from Daniel Bernoulli written in 1762. Both facts are consistent with Buffon's having continued work on the essay into 1764, when his attention would naturally have been drawn to Bayes's paper.

12. It is possible to find further parallels as well. Price cautions that "it should be carefully remembered that these deductions suppose a previous total ignorance of nature"; although such a condition is not initially stipulated by Buffon, later on in Section 8 the caveat mysteriously appears: "toutes les fois qu'un effet, dont nous ignorons absolument la cause, arrive de la même façon . . ."

13. Buffon's discussion in Section 9 of the *Essai* confirms that he believes the correct exponent to be the number of occurrences.

14. Todhunter was either being much more diplomatic or much more sarcastic when he wrote: "In the case of a writer like Laplace who agrees with his predecessors, not in one or two points but in very many, it is of course obvious that he must have borrowed largely, and we conclude that he supposed the erudition of his contemporaries would be sufficient to prevent them from ascribing to himself more than was justly due" (1865, pp. x–xi).

15. For Laplace's citation practices in his original research papers, see Stigler (1978).

16. This is clear from Laplace's wording in the preceding passage, which discusses the rule of succession.

17. That is, the Bayes/Price formula, although correct, answers the wrong question. Buffon interprets it as giving the posterior probability that the sun will rise, whereas it really gives the posterior probability that the chance x of the sun's rising is greater than $\frac{1}{2}$. That Buffon misreads it as giving the succession probability is confirmed by his language in Section 9 of the *Essai*.

18. Venn's retraction was implicit; having first harshly criticized the testimonial literature, in a later chapter "On the Credibility of Extraordinary Stories," Venn somewhat inconsistently employs the classical formulae for purposes of illustration, and admits to errors in his treatment of the subject in the first edition (1866) of the *Logic of Chance* (Venn, 1888, Chapter 17).

19. See generally Sobel (1987). For discussion of Condorcet's analysis, see Todhunter (1865, pp. 400–406); Maistrov (1974, pp. 134–135); Pearson (1978, pp. 459–461).

20. I thank Keith Baker for a helpful discussion on the subject of the paper, and Stephen Stigler for his comments on the manuscript.

Bayes, Thomas (1764). An essay towards solving a problem in the doctrine of chances. *Philosophical Transactions of the Royal Society of London* **53**, 370–418. Reprinted in Pearson, E. S., & Kendall, M. G., (eds.), *Studies in the History of Statistics and Probability*, London: Charles Griffin, (1954), 134–156. (Page citations in the text are to this edition.)

Bertrand, J. (1889). *Calcul des probabilités*. Paris: Gauthier-Villars, (2nd ed., 1907). Reprinted by New York: Chelsea.

Buffon, Georges-Louis Leclerc, Comte de (1777). *Essai d' arithmétique morale*. In Buffon, *Supplément à l' Histoire Naturelle*, vol. 4, 46–148. Reprinted in Buffon's *Oeuvres philosophiques* (J. Piveteau, ed.), Paris (1954), 456–488. (Page citations in the text are to this edition.)

Dale, A. I. (1982). Bayes or Laplace? An examination of the origin and early applications of Bayes' theorem. *Archive for History of Exact Sciences* **27**, 23–47.

De Morgan, Augustus (1838). *An Essay on Probabilities, and on Their Application to Life Contingencies and Insurance Offices*. London: Longman, Orme, Brown, Green, & Longmans, and John Taylor.

Edwards, A. W. F. (1978). Commentary on the arguments of Thomas Bayes. *Scandinavian Journal of Statistics* **5**, 116–118.

Gouraud, Charles (1848). *Histoire de calcul des probabilités*. Libraire d'Auguste Durand, Paris.

Herschel, John F. W. (1857). *Essays from the Edinburgh and Quarterly Reviews, with Addresses and Other Pieces*. London: Longman, Brown, Green, Longmans, and Roberts.

Hume, David (1739). *A Treatise of Human Nature*. Page references are to the 2nd edition of the L. A. Selby-Bigge text, revised by P. H. Nidditch, Oxford, UK: The Clarendon Press, 1978.

Keynes, John Maynard (1921). *A Treatise on Probability*. London: Macmillan.

Laplace, Pierre Simon Marquis de (1774). Mémoire sur la probabilité des causes par les évenements. *Mémoires de l' Académie royale des sciences presentés par divers savans* 6, 621–656. Reprinted in *Oeuvres complètes de Laplace*, vol. 8 (1891), Paris: Gauthier-Villars, pp. 27–65. Translated with commentary by S. M. Stigler, *Statistical Science* **1** (1986), 359–378.

Laplace, Pierre Simon Marquis de (1812). *Théorie analytique des probabilités*. Paris: Courcier (2nd ed., 1814; 3rd ed., 1820). Page references in the text are to *Oeuvres complètes de Laplace*, vol. 7 (1886), Paris: Gauthier-Villars.

Laplace, Pierre Simon Marquis de (1814). *Essai philosophique sur les probabilités*. Courcier, Paris. Page references in the text are to *Oeuvres complètes de Laplace*, vol. 7 (1886), Paris: Gauthier-Villars. Translated 1902, as *A Philosophical Essay on Probabilities* (F. W. Truscott & F. L. Emory, trans.); translation reprinted 1951, New York: Dover.

Maistrov, L. E. (1974). *Probability Theory: A Historical Sketch*. New York: Academic Press.

Murray, F. H. (1930). Note on a scholium of Bayes. *Bulletin of the American Mathematical Society* **36**, 129–132.

Pearson, Karl (1978). *The History of Statistics in the* 17th *and* 18th *Centuries*. Edited by E. S. Pearson. New York: Macmillan.

Sobel, Jordan Howard (1987). On the evidence of testimony for miracles: a Bayesian interpretation of David Hume's analysis. *The Philosophical Quarterly* **37**, 166–186.

Stigler, Stephen M. (1978). Laplace's early work: chronology and citations. *Isis* **69**, 234–254.

Stigler, Stephen M. (1982). Thomas Bayes's Bayesian inference. *J. Roy. Statist. Soc. Series A* **145**, Part II, 250–258.

Stigler, Stephen M. (1986). *The History of Statistics*. Cambridge, Mass: Harvard University Press.

Todhunter, Isaac (1865). *A History of the Mathematical Theory of Probability from the Time of Pascal to that of Laplace*. London: Macmillan. Reprinted 1949, 1965 by New York: Chelsea Publishing Company.

Venn, John (1888). *The Logic of Chance*, 3rd ed. London: Macmillan. Reprinted by Chelsea Publishing Company, New York, 1962.

Von Wright, Georg Henrik (1941). *The Logical Problem of Induction*. Second revised edition 1957, New York: Macmillan.

Zabell, S. L. (1988). The rule of succession. *Erkenntnis*, **31**, 283–321.

4

W. E. Johnson's "Sufficientness" Postulate

How do Bayesians justify using conjugate priors on grounds other than mathematical convenience? In the 1920s the Cambridge philosopher William Ernest Johnson in effect characterized symmetric Dirichlet priors for multinomial sampling in terms of a natural and easily assessed subjective condition. Johnson's proof can be generalized to include asymmetric Dirichlet priors and those finitely exchangeable sequences with linear posterior expectation of success. Some interesting open problems that Johnson's result raises, and its historical and philosophical background, are also discussed.

Key words and phrases: W. E. Johnson, sufficientness postulate, exchangeability, Dirichlet prior, Rudolph Carnap.

1. INTRODUCTION

In 1932 a posthumously published article by the Cambridge philosopher W. E. Johnson showed how symmetric Dirichlet priors for infinitely exchangeable multinomial sequences could be characterized by a simple property termed "Johnson's sufficiency postulate" by I. J. Good (1965). (Good (1967) later shifted to the term "sufficientness" to avoid confusion with the usual statistical meaning of sufficiency.) Johnson could prove such a result, prior to the appearance of de Finetti's work on exchangeability and the representation theorem, for Johnson had himself already invented the concept of exchangeability, dubbed by him the "permutation postulate" (see Johnson, 1924, page 183). Johnson's contributions were largely overlooked by philosophers and statisticians alike until the publication of Good's 1965 monograph, which discussed and made serious use of Johnson's result.

This research was supported by Office of Naval Research Contract N00014-76-C-0475 (NR-042-267).
AMS 1980 *subject classification.* Primary 62A15; secondary 62-03, 01A60.
Reprinted with permission from *The Annual Statistics* 10, no. 4 (1982): 1091–1099.

Due perhaps in part to the posthumous nature of its publication, Johnson's proof was only sketched and contains several gaps and ambiguities; the major purpose of this paper is to present a complete version of Johnson's proof. This seems of interest both because of the result's intrinsic importance for Bayesian statistics and because the proof itself is a simple and elegant argument which requires little technical apparatus. Furthermore, it can be easily generalized to characterize both asymmetric Dirichlet priors and finitely exchangeable sequences with posterior expectation of success linear in the frequency count, and the proof below is given in this generality.

The generalization of Johnson's proof mentioned above is given in Section 2. Section 3 discusses a number of complements to the result and some open problems it raises, and Section 4 concludes with a historical note on Johnson and the reception of his work in the philosophical literature.

2. FINITE EXCHANGEABLE SEQUENCES

Let $X_1, X_2, \ldots, X_{N+1}$ be a sequence of random variables, each taking values in the set $\mathbf{t} = \{1, 2, \ldots, t\}$, $N \geq 1$ and $t \leq \infty$, such that

$$P\{X_1 = i_1, \ldots, X_N = i_N\} > 0, \quad \text{for all} \quad (i_1, \ldots, i_N) \in \mathbf{t}^N. \quad (2.1)$$

Let $\mathbf{n} = \mathbf{n}(X_1, \ldots, X_N)$ denote the t-vector of frequency counts, i.e., $\mathbf{n} = (n_1, n_2, \ldots, n_t)$, where $n_i = n_i(X_1, \ldots, X_N) = \#\{X_j = i\}$. Johnson's sufficientness postulate assumes that

$$P\{X_{N+1} = i | X_1, \ldots, X_N\} = f_i(n_i), \quad (2.2)$$

that is, the conditional probability of an outcome in the ith cell given $X_1, \ldots, X_N$ only depends on n_i, the number of outcomes in that cell previously. (Note that (2.2) is well-defined because of (2.1).) If $X_1, \ldots, X_{N+1}$ is exchangeable, $f_i(n_i) = P\{X_{N+1} = i | \mathbf{n}\} = P\{X_{N+1} = i | n_i\}$.

Lemma 2.1. *If $t > 2$ and (2.1), (2.2) hold, then there exist constants $a_i \geq 0$ and b such that for all i,*

$$f_i(n_i) = a_i + b n_i. \quad (2.3)$$

Proof. First assume $N \geq 2$. Let

$$\mathbf{n}_1 = (n_1, \ldots, n_i, \ldots, n_j, \ldots, n_k, \ldots, n_t)$$

85

be a fixed ordered partition of N, with i, j, k three fixed distinct indices such that $0 < n_i, n_j$ and $n_i, n_k < N$, and let

$$\mathbf{n}_2 = (n_1, \ldots, n_i + 1, \ldots, n_j - 1, \ldots, n_k, \ldots, n_t)$$
$$\mathbf{n}_3 = (n_1, \ldots, n_i, \ldots, n_j - 1, \ldots, n_k + 1, \ldots, n_t)$$
$$\mathbf{n}_4 = (n_1, \ldots, n_i - 1, \ldots, n_j, \ldots, n_k + 1, \ldots, n_t).$$

Note that for any $\mathbf{n}$,

$$\sum_{n_l \in n} f_l(n_l) = 1, \tag{2.4}$$

hence taking $\mathbf{n} = \mathbf{n}_1, \mathbf{n}_2, \mathbf{n}_3, \mathbf{n}_4$, we obtain

$$
\begin{aligned}
f_i(n_i + 1) - f_i(n_i) &= f_j(n_j) - f_j(n_j - 1) \\
&= f_k(n_k + 1) - f_k(n_k) \\
&= f_i(n_i) - f_i(n_i - 1). \tag{2.5}
\end{aligned}
$$

Thus

$$f_i(n_i) = a_i + bn_i,$$

where we define $a_i = f_i(0) \geq 0$ and $b = \Delta f_i(n_i)$ is independent of i (because of (2.5)).

If $N = 1$, let $c_i = f_i(1)$; it then follows from (2.4) that for any i and j, $a_i + c_j = a_j + c_i$, hence $c_i - a_i = c_j - a_j = b$. $\qquad\square$

Remark 1. If $t = 2$, Johnson's sufficientness postulate is vacuous and (2.3) need not hold; see Good (1965, page 26). Thus, in the binomial case, it is necessary to make the additional assumption of linearity. In either case ($t = 2$ or $t > 2$), Johnson's argument requires that $a_i > 0$; the next two remarks address this point and are both applied in Lemma 2.2 below.

Remark 2. If (2.1) holds for $N + 1$ as well as N, then $a_i > 0$. The reader can, if he wishes, simply replace (2.1) by this strengthened version in the sequel, and ignore the following remark on a first reading.

Remark 3. If $X_1, \ldots, X_{N+1}$ is exchangeable and (2.1) holds for N, then $a_i > 0$ if $N \geq 2$. (If $a_i = 0$ for some i, then $f_i(1) > 0$, hence $b > 0$. But if $a_i = 0$, then $f_j(N - 1) = 0$ for $j \neq i$, hence $b \leq 0$, a contradiction.) This need not hold when $N = 1$; for example, let $t = 2$ and $P(1, 1) = P(2, 2) = 1/2$. This is the reason for assuming $N_0 \geq 3$ in Theorem 2.1 below: if $N_0 = 1$ the statement is vacuous, while, if $N_0 = 2$, $k_i = 0$ can occur (unless the strengthened version of (2.1) is assumed).

Let $A = \sum_i a_i$. It follows from (2.3), (2.4) that

$$A + bN = 1, \tag{2.6}$$

hence A is finite and

$$b = (1 - A)/N. \tag{2.7}$$

Suppose $b \neq 0$. Then letting $k_i = a_i/b$ and $K = \sum k_i$, we see from (2.6) that

$$b^{-1} = N + A/b = N + K,$$

hence

$$f_i(n_i) = a_i + bn_i = \frac{k_i + n_i}{b^{-1}} = \frac{n_i + k_i}{N + K}.$$

Example 2.1. (Sampling without replacement.) Let $X_1 = x_1, \ldots, X_{N+1} = x_{N+1}$ denote a random sample drawn from a finite population with $m_i \geq 1$ members in each category i. Let $M = m_1 + \cdots + m_t$ and let $N \leq m_i$, all i. Then

$$P\{X_{N+1} \in \text{category } i \,|\, \mathbf{n}\} = \frac{m_i - n_i}{M - N} = \left(\frac{m_i}{M - N}\right) + \left(\frac{1}{N - M}\right)n_i. \tag{2.8}$$

Thus $a_i = m_i/(M - N)$ and $b = (N - M)^{-1} < 0$. Note that $k_i = -m_i$; thus k_i (and hence K) is independent of N, although a_i, A, and b are not. The next lemma states that this is always the case if, as here, the X_i are exchangeable and $b \neq 0$.

Let $a_i^{(N)}$, $b^{(N)}$, $k_i^{(N)}$, and $f_i(n_i, N)$ denote the dependence of a_i, b, k_i, and $f_i(n_i)$ on N. Thus, if (2.1) and (2.2) are satisfied for a fixed $N \geq 1$, then there exist $a_i^{(N)}$ and $b^{(N)}$ such that for all i, $f_i(n_i, N) = a_i^{(N)} + b^{(N)}n_i$. Note that $b^{(N)} = 0$ if and only if $\{X_1, \ldots, X_N\}$ and X_{N+1} are independent.

Lemma 2.2. *Let* $X_1, X_2, \ldots, X_{N+1}, X_{N+2}$ *be an exchangeable sequence of* t*-valued random variables,* $N \geq 1$ *and* $t \geq 2$*, satisfying (2.1) and (2.3) for both N and $N + 1$.*

(i) If $b^{(N)} \cdot b^{(N+1)} = 0$*, then* $b^{(N)} = b^{(N+1)} = 0$*.*
(ii) If $b^{(N)} \cdot b^{(N+1)} \neq 0$*, then* $b^{(N)} \cdot b^{(N+1)} > 0$ *and* $k_i^{(N)} = k_i^{(N+1)}$*, all i.*

Proof. (i) Choose and fix two distinct indices $i \neq j$. Let $a_i = a_i^{(N)}, a_i' = a_i^{(N+1)}, b = b^{(N)}, b' = b^{(N+1)}$, etc. Suppose $b = 0$. It follows from exchangeability that for any partition $\mathbf{n}$ of N,

$$P \{X_{N+1} = i, X_{N+2} = j \mid \mathbf{n}\} = P \{X_{N+1} = j, X_{N+2} = i \mid \mathbf{n}\}, \qquad (2.9)$$

hence

$$(a_i)(a_j' + b'n_j) = (a_j)(a_i' + b'n_i). \qquad (2.10)$$

First taking $\mathbf{n}$ in (2.10) with $n_i = 0, n_j = N$, then with $n_i = N, n_j = 0$ and subtracting, we obtain $a_i b' N = -a_j b' N$, hence $b' = 0$ (since $a_i, a_j > 0$). Similarly, if $b' = 0$ then $b = 0$. (ii) Suppose $b \cdot b' \neq 0$. Then it follows from (2.9) that for any partition $\mathbf{n}$ of N,

$$\left(\frac{n_i + k_i}{N + K} \right) \left(\frac{n_j + k_j'}{N + 1 + K'} \right) = \left(\frac{n_j + k_j}{N + K} \right) \left(\frac{n_i + k_i'}{N + 1 + K'} \right), \qquad (2.11)$$

hence

$$k_i n_j + k_j' n_i + k_i k_j' = k_i' n_j + k_j n_i + k_i' k_j. \qquad (2.12)$$

Letting $n_i = 0, n_j = N$ in (2.12), then $n_i = N, n_j = 0$ and subtracting, we obtain $k_i + k_j = k_i' + k_j'$; since i and j were arbitrary, this implies $K = K'$ and, if $t > 2$, $k_i = k_i'$ for all i. Since $a_i, a_i' > 0$, clearly b and b' must have the same sign.

Suppose $t = 2$ (so that $i = 1, j = 2$, say, and $K = k_1 + k_2$). Taking $n_i = 0$, $n_j = N$ in (2.12), we obtain $k_1(N + k_2') = k_1'(N + k_2)$, hence

$$k_1(N + K) = k_1'(N + K)$$

from which it follows (since $N + K = b^{-1} \neq 0$) that $k_1 = k_1'$, hence $k_2 = k_2'$. $\qquad \square$

Together, Lemmas 2.1 and 2.2 immediately imply the following.

Theorem 2.1. *Let $X_1, X_2, \dots, X_{N_0}(N_0 \geq 3)$ be an exchangeable sequence of t-valued random variables such that for every $N < N_0$,(i) (2.1) holds, (ii) (2.2) holds if $t > 2$ or (2.3) holds if $t = 2$. If the $\{X_j\}$ are not independent ($\Leftrightarrow b^{(1)} \neq 0$), then there exist constants $k_i \neq 0$, either all positive or all negative, such that $N + \sum k_i \neq 0$ and*

$$P \{X_{N+1} = i \mid \mathbf{n}\} = \frac{n_i + k_i}{N + \sum k_i} \qquad (2.13)$$

for every $N < N_0$, partition $\mathbf{n}$ of N, and $i \in \mathbf{t}$.

Corollary 2.1. *If $X_1, X_2, X_3, \ldots$ is an infinitely exchangeable sequence which for every $N \geq 1$, satisfies both* (i) (2.1), *and* (ii) *either* (2.2), *if $t > 2$ or* (2.3), *if $t = 2$, then $b^{(1)} \geq 0$.*

Proof. Suppose $b^{(1)} < 0$. But then $N + K = 1/b^{(N)} < 0$ for all N, which is clearly impossible. □

Corollary 2.2. *For all $N \leq N_0$, under the conditions of Theorem 2.1,*

$$P\{X_1 = i_1, X_2 = i_2, \ldots, X_N = i_N\} = \frac{\prod_{i=1}^{t} \left\{ \prod_{j=0}^{n_i - 1} (j + k_i) \right\}}{\prod_{j=0}^{N-1} (j + K)}$$

$$= \frac{\Gamma(K)}{\Gamma(N + K)} \prod_{i=1}^{t} \left\{ \frac{\Gamma(n_i + k_i)}{\Gamma(k_i)} \right\}.$$

$$(2.14)$$

Proof. It follows from the product rule for conditional probabilities that it suffices to prove $P\{X_1 = i\} = k_i/K$ for all $i \in \mathbf{t}$. But

$$P\{X_1 = i, X_2 = j\} = \{a_j^{(1)} + b^{(1)}\delta_j(i)\} P\{X_1 = i\}, \qquad (2.15)$$

where $\delta_j(i)$ is the indicator function of $\{i = j\}$. Summing over i in (2.5) gives $P\{X_2 = j\} = a_j^{(1)} + b^{(1)} P\{X_1 = j\}$, hence by exchangeability $P\{X_1 = j\} = a_j^{(1)}/(1 - b^{(1)}) = k_i/K$, since $a_j^{(1)} = k_i b^{(1)}$, $1 - b^{(1)} = A^{(1)}$ (cf. (2.6)), and $K = A^{(1)}/b^{(1)}$. □

It follows from Corollary 2.2 that $\{k_i : i \in \mathbf{t}\}$ uniquely determines $P = \mathcal{L}(X_1, X_2, \ldots, X_{N_0})$. Conversely, for every summable sequence of constants $\{k_i\}$, all of the same sign, there exists a maximal sequence of $\mathbf{t}$-valued random variables $X_1, X_2, \ldots, X_{N_0}(N_0 \leq \infty)$ such that (2.1) and (2.13) hold. The length of this sequence is determined by N^*, the largest value of N such that

$$p_{i,N} = \frac{n_i + k_i}{N + K}$$

determines a probability measure on $\mathbf{t}$, i.e., $N_0 = N^* + 1$, where

(i) if $k_i > 0$, all i, and $\sum k_i < \infty$, then $N^* = \infty$, or
(ii) if $k_i < 0$, all i, and $\sum |k_i| < \infty$, then

$$N^* = \max \{N \geq 0 : N + K < 0; N + k_i \leq 0, \text{ all } i\}.$$

Thus, if $K < 0$, N^* is the integer part of $\min\{|k_i| : i \in \mathbf{t}\}$. Hence, if $N_0 > 1$, then $t < \infty$ (since $\sum |k_i| < \infty$ implies $k_i \to 0$).

When $k_i > 0$ and $t < \infty$, the cylinder set probabilities in (2.14) coincide with those arising from the Dirichlet distribution, and the characterization referred to at the end of Section 1 follows.

3. COMPLEMENTS AND EXTENSIONS

3.1. The Symmetric Dirichlet

Johnson considered the special case where (i) f_i is independent of i, i.e., for each N, there exists a *single* function f such that

$$P\{X_{N+1} = i \mid \mathbf{n}\} = f(n_i, N) \text{ for all } i; \tag{3.1}$$

(ii) b is positive (This is the major gap in Johnson's proof. If $\{X_1, X_2, \ldots\}$ is infinitely exchangeable, but not independent, the assumption that b is positive is superfluous; see Corollary 2.1 above.)

Under these conditions $t < \infty$, $a_i \equiv a$, $k_i \equiv k > 0$, $P\{X_1 = i\} = \frac{1}{t}$,

$$P\{X_{N+1} = i \mid \mathbf{n}\} = \frac{n_i + k}{N + kt}, \tag{3.2}$$

and $X_1, \ldots, X_N$ can be extended to an infinitely exchangeable sequence, whose mixing measure dF in the de Finetti representation is the symmetric Dirichlet distribution with parameter k. Good (1965, page 25) suggests that Johnson was "unaware of the connection between the use of a flattening constant k and the symmetrical Dirichlet distribution." However, Johnson was at least aware of the connection when $k = 1$, for he wrote of his derivation of (1.4) via the combination postulate,

... I substitute for the mathematician's use of Gamma functions and the α-multiple integrals, a comparatively simple piece of algebra, and thus deduce a formula similar to the mathematician's, except that instead of for two, my theorem holds for α alternatives, primarily postulated as equiprobable, [Johnson (1932, page 418); Johnson's α corresponds to our t.]

3.2. Alternative Approaches

Let Δ_t be the probability simplex $\{p_i \geq 0, i = 1, \ldots, t : \sum p_i = 1\}$. Doksum (1974, Corollary 2.1) states in the present setting that a probability measure dF on Δ_t has a posterior distribution $dF(p_i \mid X_1, \ldots, X_n)$, which depends

90

on the sample only through the values of n_i and N, if and only if dF is Dirichlet or

(i) dF is degenerate at a point (i.e., $X_1, X_2, \ldots$ is independent);
(ii) dF concentrates on a random point (i.e., dF is supported on the extreme points $\{\delta_i(j) : i = 1, \ldots, t\}$ of Δ_t, so that (2.1) would not hold);
(iii) dF concentrates on two nonrandom points (i.e., $t = 2$ or can be taken to be so).

This is a slightly weaker result than Johnson's, which only makes the corresponding assumption about the posterior *expectations* of the p_i.

Diaconis and Ylvisaker (1979, pages 279–280) prove (using Ericson's theorem, 1969, page 323) that the beta family is the unique one allowing linear posterior expectation of success in exchangeable binomial sampling, i.e., $t = 2$ and $\{X_n\}$ infinitely exchangeable, and remark that their method may be extended to similarly characterize the Dirichlet priors in multinomial sampling. Ericson's results can even be applied in the finitely exchangeable case and permit the derivation of alternate expressions for the coefficients a_i and b of (2.3).

3.3. When Is Johnson's Postulate Inadequate?

In practical applications Johnson's sufficientness postulate, like exchangeability, may or may not be an adequate description of our state of knowledge. Johnson himself did not review his postulate as universally applicable:

the postulate adopted in a controversial kind of theorem cannot be generalized to cover all sorts of working problems; so it is the logician's business, having once formulated a specific postulate, to indicate very carefully the factual and epistemic conditions under which it has practical value. [Johnson (1932, pages 418–419).]

Jeffreys (1939, Section 3.23) briefly discusses when such conditions may hold. Good (1953, page 241; 1965, pages 26–27) remarks that the use of Johnson's postulate fails to take advantage of information contained in the "frequencies of frequencies" (often useful in sampling of species problems), and elsewhere (Good, 1967) advocates mixtures of symmetric Dirichlets as frequently providing more satisfactory initial distributions in practice.

3.4. Partition Exchangeability

If the cylinder sets $\{X_i = i_1, \ldots, X_n = i_N\}$ are identified with the functions $g: \{1, \ldots, N\} \to \{1, \ldots, t\}$, then the exchangeable probability measures P are precisely those P such that

$$P\{g \circ \pi\} = P\{g\}$$

for all g and all permutations π of $\mathbf{N} = \{1, 2, \ldots, N\}$. Equivalently, the exchangeable P's are those such that the frequencies $\mathbf{n}$ are sufficient statistics with $P\{. \mid \mathbf{n}\}$ uniform.

The rationale for exchangeability is the assumption that the index set $\mathbf{N}$ conveys no information other than serving to distinguish one element of a sample from another. In the situation envisaged by Johnson, Carnap (see Section 4 below), and others, a similar state of knowledge obtains *vis-a-vis* the index set $\mathbf{t}$ (think of the categories as colors). Then it would be reasonable to require of P that

$$P\{\pi_2 \circ g \circ \pi_1\} = P\{g\}$$

for all functions $g: \mathbf{N} \to \mathbf{t}$, and permutations π_1 of $\mathbf{N}$, π_2 of $\mathbf{t}$. Call such P's *partition-exchangeable*. The motivation for the name is the following. Let $\mathbf{a}(\mathbf{n}) = \{a_r: 0 \le r \le N\}$ denote the frequencies of the frequencies $\mathbf{n}$, i.e., $a_r = \#\{n_i = r\}$. Then P is partition-exchangeable if and only if the a_r are sufficient with $P\{. \mid \mathbf{a}(\mathbf{n})\}$ uniform, i.e. $P\{g_1\} = P\{g_2\}$ whenever $\mathbf{a}(\mathbf{n}(g_1)) = \mathbf{a}(\mathbf{n}(g_2))$. The set of partition-exchangeable probabilities is a convex set containing the symmetric Dirichlets. From this perspective the frequencies of frequencies emerge as maximally informative statistics and the mixtures of symmetric Dirichlets as partition-exchangeable.

It would be of interest to have extensions of Johnson's results to "representative functions" of the functional form $f = f(n_i, \mathbf{a}(\mathbf{n}))$; for partial results in this direction ($f = f(n_i, a_0)$), see Hintikka and Niiniluoto (1976), Kuipers (1978). It would also be of interest to have Johnson type results for Markov exchangeable and other classes of partially exchangeable sequences of random variables; cf. Diaconis and Freedman (1980) for the definition and further references; Niiniluoto (1981) for an initial attempt.

4. HISTORICAL NOTE

Johnson's results appear to have attracted little interest during his lifetime. C. D. Broad, in his review of Johnson's *Logic* (vol. 3, 1924), while favorable in his overall assessment of the book, was highly critical of the appendix on "eduction" (in which Johnson introduced the concept of exchangeability and characterized the multinomial generalization of the Bayes-Laplace prior!): "About the Appendix all I can do is, with the utmost respect to Mr. Johnson, to parody Mr. Hobbes's remark about the treatises of Milton and Salmasius:

'Very good mathematics; I have rarely seen better. And very bad probability: I have rarely seen worse.'" (Broad (1924, page 379); see generally pages 377–379.) Other than this, two of the few references to Johnson's work on the multinomial, prior to Good (1965), are passing comments in Harold Jeffreys's *Theory of Probability* (1939, Section 3.23), and Good (1953, pages 238–241). This general neglect is all the more surprising, inasmuch as Johnson could count among his students Keynes, Ramsey, and Dorothy Wrinch (one of Jeffreys's collaborators). (For Keynes's particular indebtedness to Johnson, see the former's *Treatise on Probability* (1921, pages 11 (footnote 1), 68–70, 116, 124 (footnote 2), 150–155); cf. Broad (1922, pages 72, 78–79), Passmore (1968, pages 345–346).)

It is ironical that in the decades after Johnson's death, Rudolph Carnap and his students would, unknowingly, reproduce much of Johnson's work. In 1945 Carnap introduced the function $c^*[= P\{X_{N+1} = i|\mathbf{n}\}]$ and proved that it had to have the form (1.4) under the assumption that all "structure-descriptions" [= partitions $\mathbf{n}$] were *a priori* equally likely (see Carnap, 1945; Carnap, 1950, Appendix). And just as Johnson grew uneasy with his combination postulate, so too Carnap would later introduce the family of functions $\{c_\lambda: 0 \leq \lambda \leq \infty\}$ [$= (n_i + k)/(N + kt)$, λ corresponding to our k], the so-called "continuum of inductive methods" (Carnap, 1952). But while Johnson proved that (3.2) followed from the sufficientness postulate (3.1), Carnap initially *assumed* both, although his collaborator John G. Kemeny was soon after able to show their equivalence for $t > 2$. Subsequently Carnap generalized these results, first proving (3.2) follows from a linearity assumption ((2.3)) when $t = 2$ (Carnap and Stegmüller, 1959), and later, in his last and posthumously published work on the subject, dropping the equiprobability assumption (3.1) in favor of (2.2) (Carnap, 1980, Section 19; cf. Kuipers, 1978). For the historical evolution of this aspect of Carnap's work, see Schilpp (1963, pages 74–75, 979–980); Carnap and Jeffrey (1971, pages 1–4, 223); Jeffrey (1980, pages 1–5, 103–104).

For details of Johnson's life, see Broad (1931), Braithwaite (1949); for assessments of his philosophical work, Passmore (1968, pages 135–136, 343–346), Smokler (1967), Prior (1967, page 551). In addition to his work in philosophy, Johnson wrote several papers on economics, one of which, on utility theory, is of considerable importance; all are reprinted, with brief commentary, in Baumol and Goldfeld (1968).

ACKNOWLEDGMENT

I thank Persi Diaconis and Stephen Stigler for a number of helpful comments and references. I am particularly grateful to Dr. Michael A. Halls, of

King's College Library, for locating and providing a copy of the photograph of Johnson reproduced in the original paper. The photograph may have been taken in 1902, when Johnson became a Fellow of King's College.

REFERENCES

Baumol, W. J. and Goldfeld, S. M. (1968). *Precursors in Mathematical Economics: An Anthology*. Series of Reprints of Scarce Works on Political Economy **19**, London School of Economics and Political Science, London.

Braithwaite, R. B. (1949). Johnson, William Ernest. In *Dictionary of National Biography 1931–1940*. Oxford University Press, 489–490.

Broad, C. D. (1922). Critical notice of J. M. Keynes. *A Treatise on Probability. Mind* **31** 72–85.

Broad, C. D. (1924). Mr. Johnson on the logical foundations of science. *Mind* **33** 242–269, 367–384.

Broad, C. D. (1931). William Ernest Johnson. *Proceedings of the British Academy* **17** 491–514.

Carnap, R. (1945). On inductive logic. *Philosophy of Science* **12** 72–97.

Carnap, R. (1950). *Logical Foundations of Probability*. Chicago: University of Chicago Press. Second edition 1962 (original text reprinted with minor corrections and supplementary bibliography added).

Carnap, R. (1952). *The Continuum of Inductive Methods*. Chicago: University of Chicago Press.

Carnap, R. (1980). A basic system of inductive logic, part II. In Jeffrey, 7–155.

Carnap, R. and Jeffrey, R. C. (1971). *Studies in Inductive Logic and Probability*, volume I. Berkeley: University of California Press.

Carnap, R. and Stegmüller, W. (1959). *Induktive Logik und Wahrscheinlichkeit*. Vienna: Springer-Verlag.

Diaconis, P. and Freedman, D. (1980). De Finetti's generalizations of exchangeability. In Jeffrey, 233–249.

Diaconis, P. and Ylvisaker, D. (1979). Conjugate priors for exponential families. *Ann. Statist.* **7** 269–281.

Doksum, K. (1974). Tailfree and neutral random probabilities and their posterior distributions. *Ann. Probab.* **2** 183–201.

Ericson, W. A. (1969). Subjective Bayesian models in sampling finite populations. *J. Roy. Statist. Soc. Ser. B* **31** 195–224.

Good, I. J. (1953). The population frequencies of species and the estimation of population parameters. *Biometrika* **40** 247–264.

Good, I. J. (1965). *The Estimation of Probabilities: An Essay on Modern Bayesian Methods*. Research Monograph No. 30, Cambridge, Mass.: M.I.T. Press.

Good, I. J. (1967). A Bayesian significance test for multinomial distributions. *J. Roy. Statist. Soc. Ser. B* **29** 399–431 (with discussion).

Hintikka, J. and Niiniluoto, I. (1976). An axiomatic foundation for the logic of inductive generalization. In *Formal Methods in the Methodology of Empirical Sciences* (M. Przelecki, K. Saniawski, and R. Wojcicki, eds.), Dordrecht: D. Reidel. [Reprinted in Jeffrey (1980), pages 157–181.]

Jeffrey, R. C., ed. (1980). *Studies in Inductive Logic and Probability*, volume II. Berkeley: University of California Press.

Jeffreys, H. (1939). *Theory of Probability*. Oxford, UK: Clarendon Press. Third edition, 1961.

Johnson, W. E. (1924). *Logic, Part III. The Logical Foundations of Science*. Cambridge University Press. [Reprinted 1964, New York: Dover.]

Johnson, W. E. (1932). Probability: the deductive and inductive problems. *Mind* **41** 409–423. [Appendix on pages 421–423 edited by R. B. Braithwaite.]

Keynes, J. M. (1921). *A Treatise on Probability*. London: Macmillan.

Kuipers, T. (1978). *Studies in Inductive Probability and Rational Expectation*. Synthese Library 123. Dordrecht: D. Reidel.

Niiniluoto, I. (1981). Analogy and inductive logic. *Erkenntnis* **16** 1–34.

Passmore, J. (1968). *A Hundred Years of Philosophy*, 2nd ed. New York: Penguin.

Prior, A. N. (1967). Keynes; Johnson. In *The Encyclopedia of Philosophy* (Paul Edwards, ed.) 4 550–551. New York: Macmillan and Free Press.

Schillp, P. A., ed. (1963). *The Philosophy of Rudolph Carnap*. La Salle, Ill.: Open Court.

Smokler, H. E. (1967). Johnson, William Ernest. In *The Encyclopedia of Philosophy* (Paul Edwards, ed.) 4 292–293. New York: Macmillan and Free Press.

PART TWO

Personalities

5

The Birth of the Central Limit Theorem

Abstract. De Moivre gave a simple closed form expression for the mean absolute deviation of the binomial distribution. Later authors showed that similar closed form expressions hold for many of the other classical families. We review the history of these identities and extend them to obtain summation formulas for the expectations of all polynomials orthogonal to the constants.

Key words and phrases: Binomial distribution, Stirling's formula, history of probability, Pearson curves, Stein's identity, mean absolute deviation.

1. INTRODUCTION

Let S_n denote the number of successes in n Bernoulli trials with chance p of success at each trial. Thus $P\{S_n = k\} = \binom{n}{k} p^k (1-p)^{n-k} = b(k; n, p)$. In 1730, Abraham De Moivre gave a version of the surprising formula

$$E\{|S_n - np|\} = 2\upsilon(1 - p)b(\upsilon; n, p), \qquad (1.1)$$

where υ is the unique integer such that $np < \upsilon \leq np + 1$. De Moivre's formula provides a simple closed form expression for the mean absolute deviation (MAD) or L_1 distance of a binomial variate from its mean. The identity is surprising, because the presence of the absolute value suggests that expressions for the tail sum $\sum_{k \leq np} b(k; n, p)$ might be involved, but there are no essential simplifications of such sums (e.g., Zeilberger, 1989).

Reprinted with permission from the first part of *Statistical Science* 6, no. 3 (1991): 284–302. Originally published as part of *Closed Form Summation for Classical Distributions: Variations on a Theme of De Moivre* (coauthored by Persi Diaconis). The second half of this paper has not been included in this volume.

Dividing (1.1) by n, and using the result that the modal term of a binomial tends to zero with increasing n, it follows that

$$E\left\{\left|\frac{S_n}{n} - p\right|\right\} \to 0. \qquad (1.2)$$

De Moivre noted this form of the law of large numbers and thought it could be employed to justify passing from sample frequencies to population proportions. As he put it (De Moivre, 1756, page 242):

COROLLARY. *From this it follows, that if after taking a great number of experiments, it should be perceived that the happenings and failings have been nearly in a certain proportion, such as of 2 to 1, it may safely be concluded that the probabilities of happening or failing at any one time assigned will be very near in that proportion, and that the greater the number of experiments has been, so much nearer the truth will the conjectures be that are derived from them.*

Understanding the asymptotics of (.2) in turn led De Moivre to his work on approximations to the central term of the binomial. In Section 2, we discuss this history and argue that it was De Moivre's work on this problem that ultimately led to his proof of the normal approximation to the binomial.

De Moivre's formula is at once easy enough to derive that many people have subsequently rediscovered it, but also hard enough to have often been considered worth publishing, varying and generalizing. In Section 3, we review these later results and note several applications: one to bounding binomial tail sums, one to the Bernstein polynomial version of the Weierstrass approximation theorem and one to proving the monotonicity of convergence in (1.2).

In the second half of this article (omitted from this edition), we offer a generalization along the following lines: De Moivre's result works because $\sum_a^b (k - np)b(k; n, p)$ can be summed in closed form for any a and b. The function $x - np$ is the first orthogonal polynomial for the binomial distribution. We show that in fact *all* orthogonal polynomials (except the zeroth) admit similar closed form summation. The same result holds for many of the other standard families (normal, gamma, beta and Poisson). There are a number of interesting applications of these results that we discuss, and in particular, there is a surprising connection with Stein's characterization of the normal and other classical distributions.

De Moivre's formula arose out of his attempt to answer a question of Sir Alexander Cuming. Cuming was a colorful character whose life is discussed in a concluding postscript.

2. CUMING'S PROBLEM AND DE MOIVRE'S L_1 LIMIT THEOREM

Abraham De Moivre (1667–1754) wrote one of the first great books on probability, *The Doctrine of Chances*. First published in 1718, with important new editions in 1738 and 1756, it contains scores of important results, many in essentially their modern formulation. Most of the problems considered by De Moivre concern questions that arise naturally in the gambling context. Problem 72 of the third edition struck us somewhat differently:

> A and B playing together, and having an equal number of Chances to win one Game, engage to a Spectator S that after an even number of Games n is over, the Winner shall give him as many Pieces as he wins Games over and above one half the number of Games played, it is demanded how the Expectation of S is to be determined.

In a modern notation, De Moivre is asking for the expectation $E\{|S_n - n/2|\}$. In *The Doctrine of Chances*, De Moivre states that the answer to the question is $(n/2)E/2^n$, where E is the middle term of the binomial expansion of $(1 + 1)^n$, that is, $\binom{n}{n/2}$. De Moivre illustrates this result for the case $n = 6$ (when $E = 20$ and the expectation is $15/16$).

Problem 73 of *The Doctrine of Chances* then gives equation (1.1) for general values of p (De Moivre worked with rational numbers). At the conclusion of Problem 73, De Moivre gives the Corollary quoted earlier. Immediately following this De Moivre moves on to the central limit theorem.

We were intrigued by De Moivre's formula. Where had it come from? Problem 73, where it appears, is scarcely a question of natural interest to the gamblers De Moivre might have spoken to, unlike most of the preceding questions discussed in the *Doctrine of Chances*. And where had it gone? Its statement is certainly not one of the standard identities one learns today.

2.1. The Problem of Sir Alexander Cuming

Neither the problem nor the formula appear in the 1718 edition of *The Doctrine of Chances*. They are first mentioned by De Moivre in his *Miscellanea Analytica* of 1730, a Latin work summarizing his mathematical research over the preceding decade (De Moivre, 1730). De Moivre states there (page 99) that the problem was initially posed to him in 1721 by Sir Alexander Cuming, a member of the Royal Society.

In the *Miscellanea Analytica*, De Moivre gives the solution to Cuming's problem (pages 99–101), including a proof of the formula in the symmetric case (given in Section 2.3), but he contents himself with simply stating without

Table 1.1. *Exact values of mean absolute deviation*

n	$E\lvert S_n - np\rvert$	$E\lvert S_n - np\rvert/n$
6	0.9375	0.1563
12	1.3535	0.1128
100	3.9795	0.0398
200	5.6338	0.0282
300	6.9041	0.0230
400	7.9738	0.0199
500	8.9161	0.0178
700	10.552	0.0151
800	11.280	0.0141
900	11.965	0.0133

proof the corresponding result for the asymmetric case. These two cases then appear as Problems 86 and 87 in the 1738 edition of the *Doctrine of Chances*, and Problems 72 and 73 in the 1756 edition.

As De Moivre notes in the *Doctrine of Chances* (1756, pages 240–241), the expectation of $\lvert S_n - np\rvert$ *increases* with n, but *decreases* proportionately to n; thus he obtains for $p = \frac{1}{2}$ the values in Table 1.1. (De Moivre's values for $E\lvert S_n - np\rvert$ are inaccurate in some cases (e.g., $n = 200$) in the third or fourth decimal place.)

A proof of monotonicity is given in Theorem 3 of Section 3.2. De Moivre does not give a proof in either the symmetric or asymmetric cases, and it is unclear whether he had one, or even whether he intended to assert monotonicity rather than simply limiting behavior.

Had De Moivre proceeded no further than this, his formula would have remained merely an interesting curiosity. But, as we will now show, De Moivre's work on Cuming's problem led *directly* to his later breakthrough on the normal approximation to the binomial and here, too, the enigmatic Sir Alexander Cuming played a brief, but vital, role.

2.2. " . . . the hardest Problem that can be proposed on the Subject of Chance"

After stating the Corollary quoted earlier, De Moivre noted that substantial fluctuations of S_n/n from p, even if unlikely, were still possible and that it was desirable, therefore, that "the Odds against so great a variation . . . should be assigned"; a problem which he described as "the hardest Problem that can be proposed on the Subject of Chance" (De Moivre, 1756, page 242).

But initially, perhaps precisely because he viewed the problem as being so difficult, De Moivre seems to have had little interest in working on the questions raised by Bernoulli's proof of the law of large numbers. No discussion of Bernoulli's work occurs in the first edition of the *Doctrine of Chances*; and, in its preface, De Moivre even states that, despite the urging of both Montmort and Nicholas Bernoulli that he do so, "I willing resign my share of that Task into better Hands" (De Moivre, 1718, page xiv).

What then led De Moivre to reverse himself only a few years later and take up a problem that he appears at first to have considered both difficult and unpromising? Surprisingly, it is possible to give a definitive answer to this question.

De Moivre's solution to Cuming's problem requires the numerical evaluation of the middle term of the binomial. This is a serious computational drawback, for, as De Moivre himself noted, the direct calculation of the term for large values of n (the example that he gives is $n = 10,000$) "is not possible without labor nearly immense, not to say impossible" (De Moivre, 1730, page 102).

But this did not discourage the irrepressible Sir Alexander Cuming, who seems to have had a talent for goading people into attacking problems they otherwise might not. (Our concluding postscript gives another example.) Let De Moivre tell the story himself, in a passage from the Latin text of the *Miscellanea Analytica*, which has not, to our knowledge, been commented on before (De Moivre, 1730, page 102):

Because of this, the man I praised above [*vir supra laudatus*; i.e., Cuming] asked me whether it was not possible to think of some method [*num possem methodum aliquam excogitare*] by which that term of the binomial could be determined without the trouble of multiplication or, what would come to the same thing in the end, addition of logarithms. I responded that if he would permit it, I would attempt to see what I could do in his presence, even though I had little hope of success. When he assented to this, I set to work and within the space of one hour I had very nearly arrived at the solution to the following problem [*intra spatium unius circiter horae, eò perduxi ut potuerim solutionem sequentis Problematis prope elicere*].

This problem was "to determine the coefficient of the middle term of a very large even power, or to determine the ratio which the coefficient of the middle term has to the sum of all coefficients"; and the solution to it that De Moivre found in 1721, the asymptotic approximation

$$\frac{1}{2^n}\binom{n}{n/2} = 2\frac{21}{125}\left(1 - \frac{1}{n}\right)^n / \sqrt{n-1}$$

to the central term of the binomial, was the first step on a journey that led to his discovery of the normal approximation to the binomial 12 years later in 1733 (Schneider, 1968, pages 266–275, 292–300; Stigler, 1986, pages 70–88; Hald, 1990, pages 468–495). The 1721 date for the initial discovery is confirmed by De Moivre's later statement regarding the formula, in his privately circulated note of November 12, 1733, the *Approximatio ad Summam Terminorum Binomii* $(a + b)^N$ *in Seriem Expansi* that "it is now a dozen years or more since I had found what follows" (De Moivre, 1756, page 243).

Thus De Moivre's work on Cuming's problem led him immediately to the L_1 law of large numbers for Bernoulli trials, and eventually to the normal approximation to the binomial distribution. He appears to have regarded the two as connected, the second a refinement of the first. But there is one feature about De Moivre's train of thought that is puzzling. How did he make the leap from

$$E\left\{\left|\frac{S_n}{n} - p\right|\right\} \to 0 \quad \text{to} \quad P\left\{\left|\frac{S_n}{n} - p\right| > \epsilon\right\} \to 0?$$

De Moivre certainly knew the second statement from his work on the normal approximation to the binomial, as well as from Bernoulli's earlier work on the law of large numbers. But more than 120 years would have to elapse before Chebychev's inequality would allow one to easily reach the second conclusion from the first.

Of course, the currently recognized modes of convergence were not well delineated in De Moivre's time. One can find him sliding between the weak and strong laws in several places. His statement of the corollary: "the happenings and failings have been nearly in a certain proportion," has a clear element of fluctuation in it. In contrast, even today L_1 convergence has a distant, mathematical flavor to it. It is intriguing that De Moivre seemed to give it such a direct interpretation.

2.3. De Moivre's Proof

De Moivre's proof that $E[|S_n - n/2|] = (1/2)nE/2^n$ is simple but clever, impressive if only because of the notational infirmities of his day. Since it only appears in the Latin of the *Miscellanea Analytica* and is omitted from *The Doctrine of Chances*, we reproduce the argument here.

De Moivre's Proof of Formula (1.1), Case $p = 1/2$. Let E denote the "median term" (*terminus medius*) in the expansion of $(a + b)^n$, D and F the coefficients on either side of this term, C and G the next pair on either side, and so on. Thus the terms are $\dots, A, B, C, D, E, F, G, H, K, \dots$.

The expectation of the spectator after an even number of games is

$$E \times 0 + (D + F) \times 1 + (C + G) \times 2$$
$$+ (B + H) \times 3 + (A + K) \times 4 + \cdots.$$

Because the binomial coefficients at an equal distance from either side of the middle are equal, the expectation of the spectator reduces to

$$0E + 2D + 4C + 6B + 8A + \cdots.$$

But owing to the properties of the coefficients, it follows that

$$(n + 2)D = nE$$
$$(n + 4)C = (n - 2)D$$
$$(n + 6)B = (n - 4)C$$
$$(n + 8)A = (n - 6)B$$

$$\cdots$$

Setting equal the sum of the two columns then yields

$$nD + nC + nB + nA + \cdots + 2D + 4C + 6B + 8A \cdots$$
$$= nE + nD + nC + nB + nA + \cdots - 2D - 4C - 6B - 8A - \cdots.$$

Deleting equal terms from each side, and transposing the remainder, we have

$$4D + 8C + 12B + 16A + \cdots = nE$$

or

$$2D + 4C + 6B + 8A + \cdots = \frac{1}{2}nE.$$

Since the probabilities corresponding to each cofficient result from dividing by $(a + b)^n$, here $(1 + 1)^n = 2^n$, De Moivre's theorem follows. $\square$

Remark. For a mathematician of his stature, surprisingly little has been written about De Moivre. Walker's brief article in *Scripta Mathematica* (Walker, 1934) gives the primary sources for the known details of De Moivre's life; other accounts include those of Clerke (1894), David (1962, pages 161–178), Pearson (1978, pages 141–146) and the *Dictionary of Scientific Biography*.

Schneider's detailed study (Schneider, 1968) provides a comprehensive survey of De Moivre's mathematical research. During the last two decades, many books and papers have appeared on the history of probability and statistics, and a number of these provide extensive discussion and commentary on this aspect of De Moivre's work; these include most notably, the books by Stigler (1986) and Hald (1990). Other useful discussions include those of Daw and Pearson (1972), Adams (1974), Pearson (1978, pages 146–166), Hald (1984, 1988) and Daston (1988, pages 250–253).

3. LATER PROOFS, APPLICATIONS AND EXTENSIONS

3.1. Later Proofs

De Moivre did not give a proof of his expression for the MAD in the case of the asymmetrical binomial (although he must have known one). This gap was filled by Isaac Todhunter (1865, pages 182–183) who supplied a proof in his discussion of this portion of De Moivre's work.

Todhunter's proof proceeds by giving a closed form expression for a sum of terms in the expectation, where the sum is taken from the outside in. We abstract the key identity in modern notation.

Lemma 1. *(Todhunter's Formula). For all integers* $0 \leq \alpha < \beta \leq n$,

$$\sum_{k=\alpha}^{\beta}(k - np)b(k; n, p) = \alpha q b(\alpha; n, p) - (n - \beta)pb(\beta; n, p).$$

Proof. Because $p + q = 1$,

$$\sum_{k=\alpha}^{\beta}(k - np)b(k; n, p) = \sum_{k=\alpha}^{\beta}\{kq - (n - k)p\}b(k; n, p)$$

$$= \sum_{k=\alpha}^{\beta}kq b(k; n, p) - \sum_{k=\alpha}^{\beta}(n - k)pb(k; n, p).$$

But $(k + 1)q b(k + 1; n, p) = (n - k)pb(k; n, p)$; thus every term in the first sum (except the lead term) is canceled by the preceding term in the second sum, and the lemma follows. $\square$

106

We know of no proof for the $p \neq 1/2$ case prior to that given in Todhunter's book. Todhunter had an encyclopedic knowledge of the literature, and it would have been consistent with his usual practice to mention further work on the subject if it existed. He (in effect) proved his formula by induction.

Todhunter assumed, however, as did De Moivre, that np is integral (although his proof does not really require this); and this restriction can also be found in Bertrand (1889, pages 82–83). Bertrand noted that if $q = 1 - p$ and

$$F(p, q) =: \sum_{k > np} \binom{n}{k} p^k q^{n-k},$$

then the mean absolute deviation could be expressed as $2pq\{\frac{\partial F}{\partial p} - \frac{\partial F}{\partial q}\}$, and that term-by-term cancellation then leads to De Moivre's formula. The first discussion we know of giving the general formula without any restriction is in Poincaré's book (1896, pages 56–60; 1912, pages 79–83): if υ is the first integer greater than np, then the mean absolute deviation is given by $2\upsilon q b(\upsilon; n, p)$. Poincaré's derivation is based on Bertrand's but is a curiously fussy attempt to fill what he apparently viewed as logical lacunae in Bertrand's proof. The derivation later appears in Uspensky's book as a problem (Uspensky, 1937, pages 176–177), possibly by the route Poincaré (1896) → Czuber (1914, pages 146–147) → Uspensky (1937).

De Moivre's identity has been rediscovered many times since. Frisch (1924, page 161) gives the Todhunter formula and deduces the binomial MAD formula as an immediate consequence. This did not stem the flow of rediscovery, however. In 1930, Gruder (1930) rediscovered Todhunter's formula, and in 1957 Johnson, citing Gruder, noted its application to the binomial MAD. Johnson's (1957) article triggered a series of generalizations. The MAD formula was also published in Frame (1945). None of these authors connected the identity to the law of large numbers so it remained a curious fact.

Remark. The formula for the mean absolute deviation of the binomial distribution can be expressed in several equivalent forms which are found in the literature. If υ is the least integer greater than np and $Y_{n,p}$ is the central term in the expansion of $(p + q)^n$, then the mean absolute derivation equals

$2\upsilon q b(\upsilon; n, p)$ (Poincaré, 1896; Frisch, 1924; Feller, 1968)

$= 2npq b(\upsilon - 1; n - 1, p)$ (Uspensky, 1937)

$= 2npq Y_{n-1}$ (Frame, 1945)

$= 2\upsilon \binom{n}{\upsilon} p^\upsilon q^{n-\upsilon+1}$ (Johnson, 1957).

In his solution to Problem 73, De Moivre states that one should use the binomial term $b(j; n, p)$ for which $j/(n - j) = p/(1 - p)$; since this is equivalent to taking $j = np$, the solution tacitly assumes that np is integral. In this case $b(j; n, p) = b(j; n - 1, p)$ and $j = \upsilon - 1$, hence

$$2npqb(j; n, p) = 2npqb(\upsilon - 1; n - 1, p);$$

thus the formula given by De Moivre agrees with the second of the standard forms.

3.2. Applications

Application 1. As a first application we give a binomial version of the Mills ratio for binomial tail probabilities.

Theorem 1. *For* $\alpha > np, n \geq 1$ *and* $p \in (0, 1)$,

$$\frac{\alpha}{n} \leq \frac{1}{b(\alpha; n, p)} \sum_{k=\alpha}^{n} b(k; n, p) \leq \frac{\alpha(1 - p)}{\alpha - np}.$$

Proof. For the upper bound, use Lemma 1 to see that

$$\alpha \sum_{k=\alpha}^{n} b(k; n, p) \leq \sum_{k=\alpha}^{n} kb(k; n, p)$$

$$= np \sum_{k=\alpha}^{n} b(k; n, p) + \alpha q b(\alpha; n, p).$$

The lower bound follows similarly. □

Remark. The upper bound is given in Feller (1968, page 151). Feller gives a much cruder lower bound. Slightly stronger results follow from Markov's continued fraction approach, see Uspensky (1937, pages 52–56). As usual, this bound is poorest when α is close to np. For example, when $p = 1/2$, and $\alpha = [n/2] + 1$, the ratio is of order $\sqrt{n}$ while the lower bound is approximately $1/2$ and the upper bound is approximately $n/4$. The bound is useful in the tails. Similar bounds follow for other families which admit a closed form expression for the mean absolute deviation.

Application 2. De Moivre's formula allows a simple evaluation of the error term in the Bernstein polynomial approximation to a continuous function. Lorentz (1986) or Feller (1971, Chapter 8) give the background to Bernstein's approach.

Let f be a continuous function on [0, 1]. Bernstein's proof of the Weierstrass approximation theorem approximates $f(x)$ by the Bernstein polynomial

$$B(x) = \sum_{i=0}^{n} f\left(\frac{i}{n}\right)\binom{n}{i}x^i(1-x)^{n-i}.$$

The quality of approximation is often measured in terms of the modulus of continuity:

$$\omega_f(\delta) = \sup_{|x-y|\leq\delta} |f(y) - f(x)|.$$

With this notation, we can state the following theorem.

Theorem 2. *Let f be a continuous function on the unit interval. Then for any* $x \in [0, 1]$

$$|f(x) - B(x)| \leq \omega_f\left(\frac{1}{\sqrt{n}}\right)\left(1 + \frac{2\upsilon(1-x)}{\sqrt{n}}b(\upsilon; n, x)\right)$$

with $nx < \upsilon < nx + 1$.

Proof. Clearly

$$|f(x) - B(x)|$$
$$\leq \sum_{i=0}^{n}\left|f(x) - f\left(\frac{i}{n}\right)\right|\binom{n}{i}x^i(1-x)^{n-i}.$$

For any $\delta \in (0,1)$, dividing the interval between x and i/n into subintervals of length smaller than δ shows

$$|f(x) - f(i/n)| \leq \omega_f(\delta)\left(1 + \frac{|x - i/n|}{\delta}\right).$$

Using this and De Moivre's formula gives the theorem, taking $\delta = 1/\sqrt{n}$. $\square$

Remark. (1) Lorentz (1986, pages 20, 21) gives $|f(x) - B(x)| \leq \frac{5}{4}\omega_f(1/\sqrt{n})$. Lorentz shows that the function $f(x) = |x - \frac{1}{2}|$ has $|f(x) - B(x)| \geq \frac{1}{2}\omega_f(1/\sqrt{n})$ so the $1/\sqrt{n}$ rate is best possible.

(2) To get a uniform asymptotic bound from Theorem 2, suppose n is odd. Then Blyth (1980) shows that the mean absolute deviation (given by formula (1.1)) is largest for $p = \frac{1}{2}$. The upper bound in theorem 2 becomes

$$|f(x) - B(x)| \leq \omega_f\left(\frac{1}{\sqrt{n}}\right)\cdot\left(1 + \frac{(n+1)}{2\sqrt{n}}b\left(\frac{n+1}{2}; n, \frac{1}{2}\right)\right).$$

By Stirling's formula the right hand side is asymptotic to $\omega_f(\frac{1}{\sqrt{n}})(12 + \frac{1}{\sqrt{2\pi}})$.

(3) Bernstein polynomials are useful in Bayesian statistics because of their interpretation as mixtures of beta distributions (see Dallal and Hall, 1983; Diaconis and Ylvisaker, 1985). The identities for other families presented in Section 4 can be employed to give similar bounds for mixtures of other families of conjugate priors.

Application 3. As a final application, we apply the general form of De Moivre's formula (1.1) to show that the MAD of S_n is increasing in n, but that the MAD of S_n/n is decreasing in n. For S_n, let $v_n = [np + 1] = [np] + 1$, so that $np < v_n \leq np + 1$.

Theorem 3. Let $S_n \sim B(n, p)$ and $M_n =: E[|S_n - np|]$. If p is fixed, then for every $n \geq 1$,

$$M_n \leq M_{n+1}, \text{ with equality precisely when } (n + 1)p \text{ is integral;} \quad (3.1)$$

$$\frac{M_n}{n} \geq \frac{M_{n+1}}{n + 1}, \text{ with equality precisely when } np \text{ is integral.} \quad (3.2)$$

Proof. It is necessary to consider two cases.

Case 1. $v_n = v_{n+1}$. Then by the general form of De Moivre's formula

$$\frac{M_{n+1}}{M_n} = \frac{(n + 1)q}{n + 1 - v_n}$$

and

$$\frac{M_{n+1}/(n + 1)}{M_n/n} = \frac{nq}{n + 1 - v_n}.$$

But $(n + 1)p < [(n + 1)p + 1] = v_{n+1} = v_n$, hence $n + 1 - v_n < (n + 1)q$, so that $M_{n+1}/M_n > 1$. Similarly, $v_n \leq np + 1$, hence $nq \leq n + 1 - v_n$, and inequality (3.2) follows, with equality if and only if $np + 1$, hence np is integral.

Case 2. $v_n < v_{n+1}$. In this case, by De Moivre's formula,

$$\frac{M_{n+1}}{M_n} = \frac{(n + 1)p}{v_n}$$

and

$$\frac{M_{n+1}/(n + 1)}{M_n/n} = \frac{np}{v_n}.$$

Since $v_n < v_{n+1}$, clearly $v_n = v_{n+1} - 1 = [(n + 1)p] \leq (n + 1)p$, and inequality (3.1) follows, with equality if and only if $(n + 1)p$ is integral. Since $np < v_n$, inequality (3.2) follows immediately, and the inequality is strict.

110

Since np integral implies $\upsilon_n = \upsilon_{n+1}$, and $(n+1)p$ integral implies $\upsilon_n < \upsilon_{n+1}$, the theorem follows. $\qquad\square$

Remark. De Moivre's formula can be applied outside the realm of limit theorems. In a charming article, Blyth (1980) notes that the closed form expansion for the MAD has a number of interesting applications. If S_n is a binomial random variable with parameters n and p, the deviation $E|S_n/n - p|$ represents the risk of the maximum likelihood estimator under absolute value loss. As p varies between 0 and $\frac{1}{2}$, the risk is roughly monotone but, if $n = 4$, $p = \frac{1}{4}$, the estimate does better than for nearby values of p. Lehmann (1983, page 58) gives De Moivre's identity with Blyth's application.

3.3. Extensions to Other Families

De Moivre's identity can be stated approximately thus: For a binomial variate, the mean absolute deviation equals twice the variance times the density at the mode. It is natural to inquire whether such a simple relationship exists between the variance σ^2 and the mean absolute deviation μ_1 for families other than the binomial. This simple question appears to have been first asked and answered in 1923 by Ladislaus von Bortkiewicz. If $f(x)$ is the density function of a continuous distribution with expectation μ, von Bortkiewicz showed that the ratio $R =: \mu_1/2\sigma^2 f(\mu)$ is unity for the gamma ("De Forestsche"), normal ("Gaussche"), chi-squared ("Helmertsche") and exponential ("zufälligen Abstände massgebende") distributions ("Fehlergesetz"); while it is $(\alpha + \beta + 1)/(\alpha + \beta)$ for the beta distribution ("Pearsonsche Fehlergesetz") with parameters α and β.

Shortly after von Bortkiewicz's paper appeared, Karl Pearson noted that the continuous examples considered by von Bortkiewicz could be treated in a unified fashion by observing that they were all members of the Pearson family of curves (Pearson, 1924). If $f(x)$ is the density function of a continuous distribution, then $f(x)$ is a member of this family if it satisfies the differential equation

$$\frac{f'(x)}{f(x)} = \frac{x+a}{b_0 + b_1 x + b_2 x^2}. \tag{3.3}$$

Then, letting $p(x) = b_0 + b_1 x + b_2 x^2$, it follows that

$$(fp)'(x) = f(x)\{(1 + 2b_2)x + (a + b_1)\}.$$

If

$$b_2 \neq -\frac{1}{2}, \quad \text{and} \quad f(x)p(x) \to 0 \text{ as } x \to \pm\infty, \qquad (3.4)$$

then integrating from $-\infty$ to ∞ yields

$$\mu = -\frac{a + b_1}{1 + 2b_2},$$

so that

$$f(x)\{x - \mu\} = \frac{(fp)'(x)}{1 + 2b_2}$$

and

$$\int_{-\infty}^{t} (x - \mu)f(x)dx = \frac{f(t)p(t)}{1 + 2b_2}. \qquad (3.5)$$

This gives the following result.

Proposition 1. *If f is a density from the Pearson family (3.3) with mean μ and (3.4) is satisfied, then*

$$\int_{-\infty}^{\infty} |x - \mu| f(x)dx = -2f(\mu)\left\{\frac{b_0 + b_1\mu + b_2\mu^2}{1 + 2b_2}\right\}.$$

Remark. If $\beta_1 =: \mu_3^2/\mu_2^3$ and $\beta_2 =: \mu_4/\mu_2^2$ denote the coefficients of skewness and kurtosis, then, as Pearson showed, this last expression may be reexpressed as

$$\mu_1 = C2\sigma^2 f(\mu) \quad \text{where} \quad C = \frac{4\beta_2 - 3\beta_1}{6(\beta_2 - \beta_1 - 1)}.$$

The constant $C = 1 \Leftrightarrow 2\beta_2 - 3\beta_1 - 6 = 0$, which is the case when the underlying distribution is normal or Type 3 (gamma). We give further results for Pearson curves in the next section [omitted].

Just as with De Moivre's calculation of the MAD for the binomial, the von Bortkiewicz-Pearson formulas were promptly forgotten and later rediscovered. Ironically, this would happen in Pearson's own journal. After the appearance in 1957 of Johnson's *Biometrika* paper on the binomial, a series of further papers appeared over the next decade which in turn rediscovered the results of von Bortkiewicz and Pearson: Ramasubban (1958) in the case of the Poisson distribution and Kamat (1965, 1966a) in the case of the Pearson family; see also the articles by Johnson (1958) Bardwell (1960) and Kamat (1966b).

112

4. STIRLING AND CUMING

In the *Miscellanea Analytica,* De Moivre states that Problem 72 in the *Doctrine of Chances* had been originally posed to him in 1721 by Alexander Cuming, whom he describes as an illustrious man *(vir clarissimus)* and a member of the Royal Society *(Cum aliquando labenta Anno 1721, Vir clarissimus Alex. Cuming Eq. Au. Regiae Societatis Socius, quaestionem infra subjectum mihi proposuisset, solutionem prolematis ei postero die tradideram).*

Thus, we have argued, Cuming was responsible for instigating a line of investigation on De Moivre's part that ultimately led to his discovery of the normal approximation to the binomial. But curiously, Cuming was also directly responsible for James Stirling's discovery of the asymptotic series for log(n!).

At some point prior to the publication of the *Miscellanea Analytica,* De Moivre discovered that Stirling had also made important discoveries concerning the asymptotic behavior of the middle term of the binomial distribution. Stirling and De Moivre were on good terms, and De Moivre, while obviously wishing to establish that he had been the first to make the discovery, was also clearly anxious to avoid an unpleasant priority dispute (at least two of which he had been embroiled in earlier in his career). And thus, as De Moivre tells us in the *Miscellanea Analytica* (1730, page 170),

As soon as [Stirling] communicated this solution to me, I asked him to prepare a short description of it for publication, to which he kindly assented, and he generously undertook to explain it at some length, which he did in the letter which I now append.

De Moivre then gave the full text (in Latin) of Stirling's letter, dated 19 June 1729. Stirling wrote:

About four years ago [i.e., 1725], I informed the distinguished Alexander Cuming that the problems of interpolation and summation of series, and other such matters of that type, which did not fall under the ordinary categories of analysis, could be solved by the differential method of Newton; this illustrious man responded that he doubted whether the problem solved by you several years earlier, concerning the behavior of the middle term of any power of the binomial, could be solved by differentials. I then, prompted by curiosity and feeling confident that I would do something that would please a mathematician of very great merit [i.e., De Moivre], took on the same problem; and I confess that difficulties arose which prevented me from quickly arriving at an answer, but I do not regret the labor if I shall nonetheless have achieved a solution so approved by you that you would see fit to insert it in your own writings. Now this is how I did it.

Stirling then went on to give, at considerable length, an illustration of his solution, but did not derive it, because "it will be described in a tract shortly to appear, concerning the interpolation and summation of series, that I am writing".

This promised book was Stirling's *Methodus Differentialis* of 1730 (which thus appeared in the same year as De Moivre's *Miscellanea Analytica),* one of the first great works on numerical analysis. In his preface, Stirling again acknowledged the crucial role of Cuming:

The problem of the discovery of the middle term of a very high power of the binomial had been solved by De Moivre several years before I had accomplished the same thing. *It is improbable that I would have thought about it up to the present day* had it not been suggested by that eminent gentleman, the most learned Alexander Cuming, who indicated that he very much doubted whether it could be solved by Newton's differential method. [Stirling, 1730, Preface; emphasis added.]

Thus Alexander Cuming appears to have played, for De Moivre and Stirling, a role similar to that of the Chevalier de Meré for Pascal and Fermat. Who was he?

5. THE QUEST FOR CUMING

At this remove of time, the question can only be partially answered, but the story that emerges is a strange and curious one, a wholly unexpected coda to an otherwise straightforward episode in the history of mathematics.

The British *Dictionary of National Biography* tells us that Cuming was a Scottish baronet, born about 1690, who briefly served in the Scottish bar (from 1714 to 1718) and then left it, under obscure but possibly disreputable circumstances. Shortly after, Cuming surfaces in London, where he was elected a Fellow of the Royal Society of London on June 30, 1720, the year before that in which De Moivre says Cuming posed his problem. The *DNB* does not indicate the reason for Cuming's election, and there is little if any indication of serious scientific output on his part. (No papers by him appear, for example, in the *Philosophical Transactions of the Royal Society of London.* This was not unusual, however, at the time; prior to a 19th century reform, members of the aristocracy could become members of the Royal Society simply by paying an annual fee.)

During the next decade, Cuming seems to have taken on the role of intellectual go-between (see Tweedie, 1922, pages 93 and 201). Cuming's chief claim to fame, however, lies in an entirely different direction. In 1729 he undertook an expedition to the Cherokee Mountains in Georgia, several years prior to the time the first settlers went there, led by James Oglethorp, in 1734. Appointed a chief by the Cherokees, Cuming returned with seven of their number to England, presenting them to King George II in an audience at Windsor Castle on June 18, 1730. Before returning, an "Agreement of Peace and Friendship" was drawn up by Cuming and signed by the chiefs, which agreement, as the 19th century *DNB* so charmingly puts it, "was the means of keeping the Cherokees our firm allies in our subsequent wars with the French and American colonists".

This was Sir Alexander's status in 1730, when De Moivre refers to him as an illustrious man and a member of the Royal Society; both conditions, unfortunately, were purely temporary. For the surprising *denouement* to Sir Alexander's career, we quote the narrative of the *DNB:*

By this time some reports seriously affecting Cuming's character had reached England. In a letter from South Carolina, bearing date 12 June 1730, . . . he is directly accused of having defrauded the settlers of large sums of money and other property by means of fictitious promissory notes. He does not seem to have made any answer to these charges, which, if true, would explain his subsequent ill-success and poverty. The government turned a deaf ear to all his proposals, which included schemes for paying off eighty millions of the national debt by settling three million Jewish families in the Cherokee mountains to cultivate the land, and for relieving our American colonies from taxation by establishing numerous banks and a local currency. Being now deeply in debt, he turned to alchemy, and attempted experiments on the transmutation of metals.

Fantastic as Cuming's alleged schemes might seem, they were of a type not new to the governments of his day. A decade earlier, thousands had lost fortunes in England and France with the bursting of the South Sea and Mississippi "bubbles."

For Cuming it was all downhill from here. A few years later, in 1737, the law finally caught up with him, and he was confined to Fleet prison, remaining there perhaps continuously until 1766, when he was moved to the Charterhouse (a hospital for the poor), where he remained until his death on August 23, 1775. He had been expelled from the Royal Society on June 9, 1757 for nonpayment of the annual fee, and when his son, also named Alexander, died some time prior to 1796, the Cuming baronetcy became extinct. By 1738, when the second edition of De Moivre's *Doctrine of Chances* appeared, association with the Cuming name had clearly become an embarrassment, and unlike the corresponding passage in the *Miscellanea Analytica*, no mention of Cuming appears when De Moivre discusses the problem Cuming had posed to him.

Thus Cuming's life in outline. Nevertheless, there remain tantalizing and unanswered questions. The account in the *Dictionary of National Biography* appears largely based on an article by H. Barr Tomkins (1878). Tomkins's article several times quotes a manuscript written by Cuming while in prison (see also Drake, 1872), and this manuscript is presumably the ultimate source for the curious schemes mentioned by the *DNB*. But although they are there presented as serious proposals, at the time that Cuming wrote the manuscript his mind appears to have been substantially deranged for several years, and the evidentiary value of the manuscript is questionable.

ACKNOWLEDGMENTS

We thank Richard Askey, David Bellhouse, Daniel Garrison, George Gasper, Ian Johnstone, Charles Stein, Steve Stigler and Gérard Letac for their comments as our work progressed. Research supported by NSF Grant DMS-89-05874.

REFERENCES

Adams, W. J. (1974). *The Life and Times of the Central Limit Theorem.* New York: Kaedmon.

Bardwell, G. E. (1960). On certain characteristics of some discrete distributions. *Biometrika* **47** 473–475.

Bertrand, J. (1889). *Calcul des probabilités.* Paris: Gauthier-Villars.

Blyth, C. R. (1980). Expected absolute error of the usual estimator of the binomial parameter. *Amer. Statist.* **34** 155–157.

Clerke, A. M. (1894). Moivre, Abraham de. *Dictionary of National Biography* **38** 116–117.

Czuber, E. (1914). *Wahrscheinlichkeitsrechnung.* Leipzig: Teubner.

Dallal, S. and Hall, W. (1983). Approximating priors by mixtures of conjugate priors. *J. Roy. Statist. Soc. Ser. B* **45** 278–286.

Daston, L. (1988). *Classical Probability in the Enlightenment.* Princeton, NJ: Princeton Univ. Press.

David, F. N. (1962). *Games, Gods, and Gambling.* New York: Hafner.

Daw, R. H. and Pearson, E. S. (1972). Abraham de Moivre's 1733 derivation of the normal curve: A bibliographical note. *Biometrika* **59** 677–680.

De Moivre, A. (1718). *The Doctrine of Chances: or, A Method of Calculating the Probabilities of Events in Play,* 1st ed. London: A. Millar. (2nd ed. 1738; 3rd ed. 1756.)

De Moivre, A. (1730). *Miscellanea Analytica de Seriebus et Quadraturis.* London: J. Tonson and J. Watts.

Diaconis, P. and Ylvisaker, D. (1985). Quantifying prior opinion. In *Bayesian Statistics 2* (J. M. Bernardo, M. H. DeGroot, D. V. Lindley and A. F. M. Smith, eds.) 133–156. Amsterdam: North-Holland.

Drake, S. G. (1872). *Early History of Georgia, Embracing the Embassy of Sir Alexander Cuming to the Country of the Cherokees, in the Year 1730*. Boston: David Clapp and Son.

Feller, W. (1968). *An Introduction to Probability and Its Applications* 1, 3rd ed. New York: Wiley.

Feller, W. (1971). *An Introduction to Probability and Its Applications* 2, 2nd ed. New York: Wiley.

Frame, J. S. (1945). Mean deviation of the binomial distribution. *Amer. Math. Monthly* **52** 377–379.

Frisch, R. (1924). Solution d'un problème du calcul des probabilités. *Skandinavisk Aktuarietidskrift* **7** 153–174.

Gruder, O. (1930). *9th International Congress of Actuaries* **2** 222.

Hald, A. (1984). Commentary on "De Mensura Sortis." *Internat. Statist. Rev.* **52** 229–236.

Hald, A. (1988). On de Moivre's solutions of the problem of duration of play, 1708–1718. *Arch. Hist. Exact Sci.* **38** 109–134.

Hald, A. (1990). *A History of Probability and Statistics and Their Applications before 1750*. New York: Wiley.

Johnson, N. L. (1957). A note on the mean deviation of the binomial distribution. *Biometrika* **44** 532–533.

Johnson, N. L. (1958). The mean deviation with special reference to samples from a Pearson type III population. *Biometrika* **45** 478–483.

Kamat, A. R. (1965). A property of the mean deviation for a class of continuous distributions. *Biometrika* **52** 288–9.

Kamat, A. R. (1966a). A property of the mean deviation for the Pearson type distributions. *Biometrika* **53** 287–289.

Kamat, A. R. (1966b). A generalization of Johnson's property of the mean deviation for a class of discrete distributions. *Biometrika* **53** 285–287.

Lehmann, E. L. (1983). *Theory of Point Estimation*. New York: Wiley.

Lorentz, G. G. (1986). *Bernstein Polynomials*, 2nd ed. New York: Chelsea.

Pearson, K. (1924). On the mean error of frequency distributions. *Biometrika* **16** 198–200.

Pearson, K. (1978). *The History of Statistics in the 17th and 18th Centuries*. New York: Macmillan.

Poincaré, H. (1896). *Calcul des Probabilités*, 1st ed. Paris: Georges Carré. (2nd ed. 1912, Paris: Gauthier-Villars.)

Ramasubban, T. A. (1958). The mean difference and the mean deviation of some discontinuous distributions. *Biometrika* **45** 549–556.

Schneider, I. (1968). Der Mathematiker Abraham de Moivre, 1667–1754. *Arch. Hist. Exact Sci.* **5** 177–317.

Stigler, S. (1986). *The History of Statistics*. Cambridge, MA: Harvard Univ. Press.

Stirling, J. (1730). *Methodus Differentialis*. London: Gul. Bowyer.

Todhunter, I. (1865). *A History of the Mathematical Theory of Probability*. London: Macmillan.

Tomkins, H. B. (1878). Sir Kenneth William Cuming of Culter, Baronet. *The Genealogist* **3** 1–11.

Tweedie, C. (1922). *James Stirling*. Oxford, UK: Clarendon Press.

Uspensky, J. V. (1937). *Introduction to Mathematical Probability*. New York: McGraw-Hill.

von Bortkiewicz, L. (1923). Über eine vershiedenen Fehlergesetzen gemeinsame Eigenschaft. *Sitzungsberichte der Berliner Mathematischen Gessellschaft* **22** 21–32.

Walker, H. M. (1934). Abraham de Moivre. *Scripta Mathematica*. **2** 316–333.

Zeilberger, D. (1989). A holonomic systems approach to binomial coefficient identities. Technical Report, Drexel Univ., Philadelphia, PA.

6

Ramsey, Truth, and Probability

Frank Ramsey's essay "Truth and Probability" represents the culmination of a long tradition at Cambridge of philosophical investigation into the foundations of probability and inductive inference; and in order to appreciate completely both the intellectual context within which Ramsey wrote, and the major advance that his essay represents, it is essential to have some understanding of his predecessors at Cambridge. One of the primary purposes of this paper is to give the reader some sense of that background, identifying some of the principal personalities involved and the nature of their respective contributions; the other is to discuss just how successful Ramsey was in his attempt to construct a logic of partial belief.

"Truth and Probability" has a very simple structure. The first two sections of the essay discuss the two most important rival theories concerning the nature of probability that were current in Ramsey's day, those of Venn and Keynes. The next section then presents the alternative advocated by Ramsey, the simultaneous axiomatization of utility and probability as the expression of a consistent set of preferences. The fourth section then argues the advantages of this approach; and the last section confronts the problem of inductive inference central to English philosophy since the time of Hume. The present paper has a structure parallel to Ramsey's; each section discusses the corresponding section in Ramsey's paper.

1. ELLIS AND VENN

Ramsey's essay begins by disposing of the frequentist and credibilist positions; that is, the two positions advocated by his Cambridge predecessors John Venn (in *The Logic of Chance*, 1866) and John Maynard Keynes (in his *Treatise on Probability*, 1921); and thus the two positions certain to be known to his audience. Toward the frequency theory Ramsey adopts a conciliatory tone: in common usage, probability often means frequency; frequencies afford a particularly simple example of quantities satisfying the laws of probability;

Reprinted with permission from *Theoria* 57 (1991): 211–38.

it may even be that frequency is the most important use of probability in science (although Ramsey makes it clear that he does not actually believe this). None of this however, Ramsey notes, excludes the possibility of constructing a logic of partial belief.

Thus, unlike hardline subjectivists such as Bruno de Finetti, Ramsey did not regard the frequency theory as necessarily wrong-headed; it might indeed have a legitimate scope, just one too narrow to be completely satisfactory. But in any case, it could not simply be dismissed, if only because its two most vigorous English advocates had both come from Cambridge! One of these was

1.1. Robert Leslie Ellis (25 August 1817–12 May 1859). Entered Trinity, 1836; Senior Wrangler and Fellow of Trinity, 1840.[1]

There are a number of curious parallels between the lives of Ellis and Ramsey: both attended Trinity, both died relatively young (Ellis at the age of 41, after having suffered from a debilitating illness for a decade); both had posthumous editions of their collected papers published shortly after their deaths by close friends (Walton, 1863; Braithwaite, 1931); both enjoyed an unusual combination of technical mathematical ability and acute philosophical insight; both had broad interests (Ellis, for example, translated and edited the works of Bacon, and devoted some effort to the construction of a Chinese dictionary); and both advocated philosophies of probability which marked a radical departure from the views of their day.

Today an obscure and largely forgotten figure, during the decade of the 1840s Ellis was an important member of the Cambridge mathematical community.[2] A student of George Peacock (one of the most influential English mathematicians of the period), after his election as a Fellow of Trinity in October 1840 Ellis first assisted and then replaced an ailing D. F. Gregory as editor of the *Cambridge Mathematical Journal*. Despite increasing administrative responsibilities, serving first as Moderator (i.e., principal mathematical examiner) for the University in 1844, and then Examiner (a position only slightly less demanding) in 1845, Ellis continued to contribute mathematical papers to a number of English journals until incapacitated by illness in 1849.

Ellis wrote six papers on the foundations of probability during this period (Ellis 1844a, 1844b, 1844c, 1844d, 1850, and 1854). From a modern perspective the most important is the first of these, one of the earliest statements of a purely frequentist theory of probability.[3] Ellis's primary argument in favor of the limiting frequency interpretation is a simple one: he states that whenever a person judges one event to be more likely to happen than another, introspection ("an appeal to consciousness") will reveal the concomitant "belief that on the long run it will occur more frequently." Such a claim is of course vulnerable to the simple observation that many uncertain events are unique and can be never be embedded in a suitable long run (for example, whether Disraeli or Gladstone will be Prime Minister after the next election) – unless, of course, one simply rejects outright the possibility that events of this kind can be assigned numerical probabilities at all.

This was in fact Ellis's position, and one that became increasingly common later in the century, being adopted in different guises by Boole, Venn, Peirce, Bertrand, and many others; thus by 1926 Ramsey could state with complete justification that "it is a common view that belief and other psychological variables are not measurable" (Braithwaite, 1931, p. 166). Perhaps the clearest statement of Ellis's position appears a letter to the scientist J. D. Forbes; such correspondence is often more revealing than the judicious language of a published paper:

The foundation of all the confusion is the notion that the numerical expression of a chance expresses the force of expectation, whereas it only expresses the proportion of frequency with which such and such an event occurs on the long run. From this notion that chances express something mental or subjective, is derived the assumption that the force of belief touching past events admits of numerical evaluation as well as the force of expectation touching future. If this were true, it would be a legitimate inquiry to try to assign numerical values to the force of belief in any given case. All this folly, for one cannot give it any other name, grows out of such statements as "certainty is equal to unity", and the like. It belongs to the school of Condillac and the sensationalists – they were in the ascendant when the theory of probabilities *received its present form, and there has not yet been philosophy enough to expel it* [Shairp *et al.* (1873, p. 481); emphasis added.]

"Its present form" – thus Ellis admitted that the frequentist view he favored was still very much a minority position in 1850.[4]

Ellis died in 1859 and for the moment his ideas largely died with him.[5] But a few years later, however, another and more sustained advocate of the frequentist position appeared on the scene. This was

1.2. John Venn (4 August 1834–4 April 1923). Entered Gonville and Caius, 1853; 6th Wrangler and Fellow of Gonville and Caius, 1857; College Lecturer in Moral Science, 1862; President, Gonville and Caius, 1903.

Venn is the direct successor to Ellis in his philosophical view of probability; and his book *The Logic of Chance* (1866) marks the first systematic account in English of a purely frequentist view of probability. Ian Hacking has suggested to me that Ellis's position should be seen as a natural consequence of Quetelet's influence in England. Hacking's conjecture may well be correct, but in the later case of Venn, speculation is unnecessary. *The Logic of Chance*, Venn's son tells us, "owed its inception to H. T. Buckle's well-known discussion concerning the impossibility of checking the statistical regularity of human action;" it thus represents in probability the philosophical expression of the rising tide of empiricism to be found throughout the European social sciences.[6]

Looking back, Venn thought that virtually the only person "to have expressed a just view of the nature and foundation of the rules of Probability" was Mill, in his *System of Logic*; the only originality Venn claimed for the essay he now put before the public was its "thorough working out of the Material view of Logic as applied to Probability."

With what may be called the Material view of Logic as opposed to the Formal or Conceptualist, – with that which regards it as taking cognisance of laws of things and not of the laws of our own minds in thinking about things, – I am in entire accordance. [Venn, 1866, p. x of the 1888 edition]

(Ellis had also viewed matters in much the same light, asking earlier how the theory could "be made the foundation of a real science, that is of a science relating to things as they really exist?"; and regarded it as an example of "the great controversy of philosophy; – I mean the contest between the realists and the nominalists" (Ellis, 1854).)

Venn's *Logic of Chance* ultimately went through three editions (1866, 1876, and 1888); the second and third editions saw many important changes in Venn's viewpoint, and it would be a useful undertaking to study these in detail.[7] The first and second editions, for example, were uncompromising in their opposition to the classical Laplacean approach (so much so that an otherwise sympathetic R. A. Fisher later felt compelled to disavow Venn's attacks on the rule of succession as unfair),[8] but the third edition displays some moderation in Venn's opposition to inverse methods, reflecting the influence of the Oxford statistician and economist Francis Ysidro Edgeworth (whose assistance is acknowledged in Venn's preface); also important at this stage was Venn's association with Francis Galton (see Stigler, 1986, pp. 291, 302 and 307).[9]

This shift, although in some respects subtle and not emphasized by Venn, was nonetheless very real. The most striking testimony to the evolution in his thinking is Venn's surprising characterization of Edgeworth (1884) as having "a view not substantially very different from mine, but expressed with a somewhat opposite emphasis" (Venn, 1888, p. 119); this was a remarkable comment given that Edgeworth's article defends the Laplacean approach to probability, arguing in part that uniform priors are justified by past empirical experience. This clearly marked a major shift in Venn's viewpoint; an impression confirmed by his comments regarding the controversy over the statistical analysis of the efficacy of Lister's antiseptic surgical procedures (Venn, 1888, pp. 187–188; see Zabell, 1989b, p. 251).

Thus Venn had made a sustained and vigorous argument in favor of what he termed the "materialist" approach to probability. His impact at the time was limited, primarily because his two most promising potential constituencies, the philosophical and statistical communities, were in one case actively hostile, and in the other largely indifferent to that viewpoint. Philosophically, Venn's materialist approach to logic was completely at odds with the particularly virulent strain of idealism then endemic in England; while the statistical profession was dominated by Edgeworth and Pearson, both of whom remained loyal to inverse methods, albeit tempered by pragmatic English common sense (see Edgeworth, 1884; Pearson, 1892). And outside of England, Venn's views were also largely neglected (with the important exception of Charles Sanders Peirce in the United States); see Kamlah (1987, p. 101).

Thus by the time Ramsey read his essay in 1926, the attractive empirical aspects of the frequency interpretation had been largely co-opted by the statisticians of the Pearson school, and fused in an eclectic fashion with the Laplacean approach to inference. For this reason Ramsey could afford to be tolerant toward frequentism in his opening comments. Just a few years earlier however, in 1921, his friend and older contemporary John Maynard Keynes had forcefully advocated a very different type of interpretation, one which could only be met head on.

2. JOHNSON AND KEYNES

Keynes's *Treatise* appeared at a time when there was a resurgence of interest at Cambridge in the foundations of probability. For three decades there had been a hiatus: from 1888 (the 3rd edition of Venn's *Logic of Chance*) to 1918, little had appeared on the foundations of the subject; and although Venn remained at Cambridge until his death in 1923 (there was one don at Gonville and Caius who in 1990 could still remember the figure of Venn on the streets of Cambridge two-thirds of a century earlier), by 1890 he had largely turned from probability and statistics. But the end of the first World War witnessed a sudden and remarkable outburst of activity centered at and around Cambridge: the years from 1918 to 1926 saw the publication of Broad's first two papers on induction (Broad, 1918 and 1920), the papers by Jeffreys and Wrinch (1919 and 1921), Keynes's *Treatise* (1921), Fisher's fundamental papers on statistical inference (Fisher, 1922 and 1925), the relevant portions of Johnson's *Logic* (1924), and the reading of Ramsey's paper *Truth and Logic* before the Cambridge Moral Sciences Club on 26 November 1926.

But such things do not occur in a vacuum, and much had happened at Cambridge in the interim. In retrospect, it is easy to point to several important factors that were responsible for this efflorescence. One of these was the presence at Cambridge of a philosopher possessing both technical competence and serious professional interest in probability theory (W. E. Johnson); the second, the revolution in English philosophy having its epicenter at Cambridge, emphasizing the use of analytic methods and formal logic (led by Moore, Russell, and Whitehead); and the third, the presence of a group of students at Cambridge (Keynes, Broad, Wrinch, and Ramsey) possessing both the technical ability and the intellectual self-assurance necessary to challenge a philosophical establishment largely hostile to quantitative argumentation. Let us consider Johnson and Keynes a little more closely.

2.1. William Ernest Johnson (23 June 1858–14 January 1931). Entered King's College, 1879; 11th Wrangler, 1882; first class honors, Moral Sciences tripos, 1883; Fellow of King's College and Sidgwick Lecturer in Moral Science, 1902–1931.

Although Johnson's father was the headmaster of a Cambridge academy, Johnson was not – in contrast to Venn – a member of what Noel Annan (1955) has termed the English "intellectual aristocracy." Perhaps in part because of this, after entering King's in 1879 on a mathematical scholarship, advancement for Johnson did not come either rapidly or easily: after lecturing on psychology and education at the Cambridge Women's Training College, Johnson held a succession of temporary positions at the University (University Teacher in the Theory of Education, 1893 to 1898; University Lecturer in Moral Science, 1896 to 1901), until 1902, when he was finally appointed Sidgwick Lecturer in Moral Science and a Fellow of King's College, where he remained until his death in 1931.[10]

I have discussed Johnson's contributions to the philosophical foundations of probability and induction in several papers (Zabell 1982, 1988, 1989a). Throughout his career Johnson experienced great difficulty in putting his ideas down in print; thus his influence at Cambridge was primarily through his lectures and personal interaction with colleagues and students. But that influence was not inconsiderable, as may be seen by turning to the work of his most famous student.

2.2. John Maynard Keynes (1883–1946). Entered Kings, 1902; First class, mathematical tripos, 1905; Fellow of Kings, 1908.

If Johnson was a son of Cambridge, Keynes *was* Cambridge; his father John Neville Keynes (1862–1949) had been a distinguished logician and economist – and for many years registrar – at the University, while the son was a student and Fellow of Kings, a member of the Apostles (elected 28 February 1903),[11] a friend of Moore, Russell, Wittgenstein, and Ramsey, and of course one of the most distinguished economists of the century.

Athough Keynes's *Treatise on Probability* (published in 1921) appeared only after his first major economic work, *The Economic Consequences of the Peace* (published in 1919), his study of the foundations of probability dates back to his student days at Cambridge, and represents his first serious intellectual interest. Initially drawn to the subject because of Moore's use of probability in the *Principia Ethica*, Keynes read a paper on the subject for the Apostles ("Ethics in relation to conduct"), perhaps as early as 1904.[12]

The *Treatise* itself was in no sense a casual effort. It began life as an unsuccessful Fellowship dissertation submitted in 1908 (but successfully revised a year later); by 1910 twenty-seven of the thirty chapters that eventually came to comprise the book had been drafted.[13] But after crossing swords with Karl Pearson later that summer in the pages of the *Journal of the Royal Statistical Society* (over a study concerning the influence of parental alcoholism), Keynes decided to add several more chapters on statistical inference, thus further delaying its publication.[14] The year 1914 once again saw Keynes at work on – and this time close to finishing – yet another revision of the manuscript, but then the outbreak of the war intervened, disrupting this as so many other human enterprises. The conflict itself and Keynes's subsequent involvement with the negotiations over the Versailles treaty prevented him from returning to the *Treatise* until 1920, when at long last he was able to find the time necessary to prepare the much-delayed work for the press.[15]

In his preface, Keynes notes that he had "been much influenced by W. E. Johnson, G. E. Moore, and Bertrand Russell;" but the nature of Keynes's debt to Johnson was a complex one. In the introductory chapter to the second, technical part of the *Treatise*, Keynes tells us:

A further occasion of diffidence and apology in introducing this Part of my Treatise arises out of the extent of my debt to Mr. W. E. Johnson. I worked out the first scheme in complete independence of his work and ignorant of the fact that he had thought, more profoundly than I had, along the same lines; I have also given the exposition its final shape with my own hands. But there was an intermediate stage, at which I submitted what I had done for his criticism, and received the benefit not only of criticism, but of his own constructive exercises. The result is that in its final form it is difficult to indicate the exact extent of my indebtedness to him. [Keynes, 1921, p. 116]

In fact Johnson, together with Whitehead, had been one of the two readers of Keynes's 1907 fellowship dissertation; and it is apparent from Keynes's statement that by 1907 Johnson had already begun his profound studies into the foundations of probability.[16]

Although his philosophical views on the nature of probability were very different from those of Venn, Keynes's *Treatise on Probability* may be viewed in many ways as the successor to Venn's *Logic of Chance*, for it was the first serious work on the foundations of probability to appear in English since the publication of the third edition of Venn's book in 1888.[17] This did not go unnoted at the time: Edgeworth's 1922 review of Keynes's *Treatise* in *Mind* is a counterpoint to his earlier review of the second, 1876 edition of Venn's *Logic* (Edgeworth, 1884); and there is a charming letter in the Caius library from Keynes to Venn (then in his 88th year), making this very point.[18] Keynes wrote:

> King's College, Cambridge
> 31 August 1921
>
> Dear Dr Venn,
>
> I have asked my publishers to send you a copy of my book *A Treatise on Probability*, which will appear this week. I send it, if I may say so, in a spirit of piety to the father of this subject in Cambridge. It is now no less than 55 years since the appearance of your first edition; yet mine is the systematic treatise on the *Logic* of the subject, next after yours, to be published from Cambridge; nor, so far as I know, has there been any such treatise in the meantime in the English language. Yours was nearly the first book on the subject that I read; and its stimulus to my mind was of course very great. So, whilst you are probably much too wise to read any more logic (as I hope I shall be in my old age), I beg your acceptance of this volume, the latest link in the very continuous chain (in spite of differences in opinion) of Cambridge thought.
>
> Yours sincerely, J M Keynes

What were those "differences in opinion"? Keynes's *Treatise* advanced a view of probability as a logical relation between propositions, sometimes but not always admitting numerical expression.[19] Unfortunately for Keynes, this view soon received a fatal challenge.

3. FRANK PLUMPTON RAMSEY

Frank Ramsey entered a world in intellectual ferment when he went up to Cambridge in 1920. Russell had been expelled from Trinity for his opposition to England's participation in the slaughter of the First World War,[20] Moore was about to become the editor of *Mind*, Johnson expounded his theories of logic and probability in the lecture hall; Johnson's *Logic*, Keynes's *Treatise*, and Wittgenstein's *Tractatus* were all about to appear.

Ramsey was admitted into this magic circle with heady rapidity: in 1920, proofreading the *Treatise* in galleys; in 1921, translating the *Tractatus* into English; in 1922 and 1923, reviewing the *Treatise* and *Tractatus* for the *Cambridge Magazine* and *Mind*; in 1923, going on a pilgrimage to visit Wittgenstein in the fastness of his rural Austrian retreat; and finally, in 1924 elected a Fellow of King's.[21]

Let us turn to Ramsey's essay. What is remarkable about it is that – notwithstanding his personal friendships with Keynes, Johnson, and Wittgenstein, and his intimate knowledge and understanding of their work – it marks a complete philosophical break with their views of the subject. Toward the frequency theory, as we have seen, Ramsey had taken a position that might be described as one of peaceful coexistence. No such accommodation was possible with Keynes's theory, however, and the second section of Ramsey's essay is in consequence a swift, skillful, and in many ways almost brutal demolition of that theory. If probability is a relation between propositions, just what is the probability relation connecting 'this is red,' and 'this is blue'? If such a relation cannot be judged against any numerical yardstick, of what use is it? The theory is so flawed, Ramsey argues, that Keynes is unable to "adhere to it consistently even in discussing first principles". Like a referee in a poorly matched boxing contest, Ramsey calls the fight before it is over: he stops his critique "not because there are not other respects in which [Keynes's theory] seems open to objection", but because it is sufficiently flawed to justify reopening the issue.

The approach that Ramsey advocates instead is very natural when viewed against the backdrop of the Moore-Russell program of philosophical *perestroika*: in order for the concept of a numerical measure of partial belief to make sense, an operational definition for it must be provided. Thus "It will not be very enlightening to be told that in such circumstances it would be rational to believe a proposition to the extent $\frac{2}{3}$, unless we know what sort of belief in it that means" (Braithwaite, 1931, p. 166); and "the degree of a belief is just like a time interval; it has no precise meaning unless we specify more exactly how it is to be measured" (p. 167). The method of measurement of our beliefs that Ramsey advances involves examining their role "*qua* basis of action;" that is, as dispositional beliefs "which would guide my action in any case to which it was relevant" (p. 172).

Ramsey in fact presents *two* closely connected systems. The first of these is virtually identical to de Finetti's own initial approach: "the old established way of measuring a person's belief is to propose a bet, and see what are the lowest odds which he will accept" (p. 172).[22] As Ramsey notes, this procedure, while "fundamentally sound", suffers from a variety of defects: the diminishing marginal utility of money; the attraction or aversion of an individual to betting; the information conveyed to that person by someone else proposing a bet.

In order to avoid these difficulties, Ramsey considers instead a somewhat more complex approach, based initially on an additive system of utilities ("the assumption that goods are additive and immediately measurable"), and the psychological assumption that "we act in the way we think most likely to realize these goods". After the consequences of these are explored, the analysis predicated on a known system of numerical utilities is discarded, and Ramsey puts forward in its place his celebrated proposal for the simultaneous axiomatization of probability and utility based on a transitive system of preferences among differing options.[23]

Given two outcomes, say α and β, if we are to pass from the qualitative assertion that α is less likely than β, to a quantitative statement about their respective probabilities, then it is necessary to provide a continuum of intermediate possibilities. Ramsey's device for achieving this was the *ethically neutral proposition;* the philosophical equivalent of tossing a coin. These are propositions p whose truth-value is never directly an object of desire in any possible world (in the sense that if p is atomic, then we are indifferent between any two possible worlds differing only in the value of p, and if p is not atomic, then its atomic constituents are each assumed to have this property).

The key step is then to specify when the occurrence of an ethically neutral proposition is equivalent to the tossing of a *fair* coin:

The subject is said to have belief of degree $\frac{1}{2}$ in such a proposition p if he has no preference between the options (1) α if p is true, β if p is false, and (2) α if p is false, β if p is true, but has a preference between α and β simply.

By successively tossing such an unbiased philosopher's coin, one can interpolate between any two options a further continuum of options, and thus establish a numerical scale for utility, uniquely determined up to affine transformation. (For a generalization of Ramsey's system resulting in a utility function unique only up to fractional linear transformation, see Jeffrey, 1983.)[24]

4. THE LOGIC OF CONSISTENCY

In the fourth section of his essay, Ramsey steps back for a moment and assesses how well the theory has done. He sees three basic advantages for it.

128

4.1. "It gives us a clear justification for the axioms of the calculus."

This was an important advance, a major advantage over systems like Keynes's (and, to cite a more recent example, Glenn Shafer's theory of belief functions), where the underlying probability relation is taken to be a primitive, undefined concept, and the axioms setting forth its basic properties are simply posited. To see what Ramsey had in mind, consider De Morgan's discussion of probability in his *Formal Logic* (1847, p. 179). From a conceptual point of view, this was one of the clearest analyses of the foundations of probability in the nineteenth century; De Morgan discusses at some length such issues as the relation between probability and belief, the numerical quantification of belief, and the additivity of the numerical measure. After concluding that partial belief is indeed capable of numerical measurement, De Morgan introduces the additivity of this measure as a "postulate"; and after lengthy discussion to impress on the reader just how far from obvious such a postulate is, he concludes that if one raises the question of its justification, "I cannot conceive any answer except that it is by an assumption of the postulate" (p. 182)!

In 1847 this might have passed muster, but after three-quarters of a century of criticism from philosophers and scientists, a more convincing response was clearly needed. It is a tribute to Ramsey's genius that *both* of the standard approaches commonly in use today to derive the axioms may be found in his essay. But equally important was a second major contribution.

4.2. "the Principle of Indifference can now be altogether dispensed with"

This was no small accomplishment. For much of the preceding century, this principle had been the subject of constant, if evolving, criticism. Boole, for example, had noted that in the case of a finite number of alternatives the principle of indifference could be applied in more than one way, to derive mutually contradictory results; Venn had ridiculed one of its primary applications, the rule of succession; von Kries had supplied examples corresponding to those of Boole for a continuum of alternatives; Bertrand had acerbically and effectively crafted a number of examples sufficiently striking that they can still be found in textbooks today.

One of the most effective parts of Keynes's *Treatise* is its criticism of the principle of indifference. Much of this was derivative, Keynes skillfully weaving together the criticisms of Boole, Venn, von Kries, and Bertrand. But it served a valuable purpose: Boole was obscure, often confused; Venn had waffled between editions; and few in the English-speaking world were familiar with the relevant French and German literature. Keynes's impact was immediate: in 1918 C. D. Broad had gingerly embraced the principle, but in the aftermath of Keynes's critique, Broad did an abrupt *volte-face*, and his 1927 paper rejected it outright.

The third and final achievement that Ramsey points to underlines the truly seminal nature of the essay: it concerns an issue that has attracted considerable interest in recent decades.

4.3. The Existence of Probable Knowledge: "I think I perceive or remember something but am not sure."

As Ramsey notes, because Keynes's theory is a "relation between the proposition in question and the things I know for certain", it cannot accommodate uncertain knowledge. Since Ramsey's theory is primarily *static*, one of consistency of partial belief at a given point in time, and does not attempt to describe the origin of those beliefs, the problem does not arise in his system. (But such problems can be analyzed within it: in recent decades the *dynamics* of how our beliefs change upon the receipt of "probable knowledge" has been extensively studied employing Richard Jeffrey's device of *probability kinematics*; see Jeffrey, 1968, Diaconis and Zabell, 1982, Jeffrey, 1988).

Ramsey did consider the question of the dynamic evolution of belief. Conditional probability is defined in terms of conditional bets; it states the odds that someone "would now bet on p, the bet only to be valid if q is true" (p. 79). This approach has since been adopted as the basis of the commonly accepted subjectivist definition (e.g., de Finetti, 1972, p. 193); but of course it does not address the relation that conditional probabilities – thus defined – may have to the actual degrees of belief one holds after the observation of an event. And here we run up against an apparent inconsistency in Ramsey's views. Initially, Ramsey notes there is no reason to automatically identify the two quantities:

the degree of belief in p given q is not the same as the degree to which [a person] would believe p, if he believed q for certain; for knowledge of q might for psychological reasons profoundly alter his whole system of beliefs.

But further on in the last section of his essay Ramsey writes:

Since an observation changes (in degree at least) my opinion about the fact observed, some of my degrees of belief after the observation are necessarily inconsistent with those I had before. We have therefore to explain how exactly the observation should modify my degrees of belief; *obviously, if p is the fact observed, my degree of belief in q after the observation should be equal to my degree of belief in q given p before*, or by the multiplication law to the quotient of my degree of belief in *pq* by my degree of belief in *p*. When my degrees of belief change in this way, we can say that they have been changed consistently by my observation.

Clearly the emphasized portion of the quotation completely ignores the profound insight of the preceding quotation; perhaps this second passage represents a portion of the text written at an earlier stage. (Ramsey's note on p. 194 makes it clear that at least some portions of the essay went through more than one draft.)

The identification of the conditional odds accepted *now* with the actual degree of belief adopted *after* is, to use Hacking's (1967) terminology, the *dynamic assumption of Bayesianism*, and the circumstances under which it is warranted has been the subject of considerable discussion in recent decades. The importance of the question is in part that the assumption plays a key role in subjectivistic analyses of inductive inference, and this brings us to the subject of the final section of Ramsey's essay.

5. THE JUSTIFICATION OF INDUCTION

Ramsey's essay thus provided an operational interpretation of subjective probability, showed how the usual axioms for probability followed as a simple consequence of that definition, and liberated the theory from the need to appeal to the principle of indifference. After Ramsey's death, Keynes paid tribute to this achievement:

The application of these ideas [regarding formal logic] to the logic of probability is very fruitful. Ramsey argues, as against the view which I had put forward, that probability is concerned not with objective relations between propositions but (in some sense) with degrees of belief, and he succeeds in showing that the calculus of probabilities simply amounts to a set of rules for ensuring that the system of degrees of belief which we hold shall be a *consistent* system. Thus the calculus of probabilities belongs to formal logic. But the basis of our degrees of belief – or the *a priori* probabilities, as they used to be called – is part of our human outfit, perhaps given us merely by natural selection, analogous to our perceptions and our memories rather than to formal logic. So far I yield to Ramsey – I think he is right. But in attempting to distinguish "rational" degrees of belief from belief in general he was not yet, I think, quite successful. It is not getting to the bottom of the principle of induction merely to say that it is a useful mental habit. (Keynes, 1951, pp. 242–244.)

This passage has usually been interpreted as a substantial recantation by Keynes of the position taken by him in his *Treatise*.[25] And rightly so; Keynes's wording, "so far I yield to Ramsey – I think he is right", is clear enough: the proper theater of operations for the theory of probability is consistent degree of belief rather than objective propositional relation. But this – although the truth – was not the whole truth for Keynes. To be "rational", our beliefs must satisfy not only the purely *static* requirement of consistency, but the *dynamic* requirement that they evolve over time in conformity with the ordinary modes of inductive inference. Despite its many other virtues, Ramsey's theory simply did not capture (or at least appear to capture) this aspect of rational belief.

This was in many ways understandable. The classical probabilistic justification for simple enumerative induction, due initially to Bayes (via his intellectual executor Price) and Laplace (in his *Essai philosophique* of 1814), employed the so-called rule of succession (see generally Zabell, 1989a); and central to the derivation of that rule was – in some guise or other – an appeal to the principle of indifference, the principle that Keynes had so effectively criticized and Ramsey so effectively eliminated.

The missing element needed was provided by Bruno de Finetti, shortly after Ramsey's death: if the possible outcomes are *exchangeable*, then (except for what might be termed cases of extreme epistemic obstinacy, which admit of precise characterization) our degrees of belief must evolve over time in inductive fashion. Earlier theories, it turned out, had been too ambitious, purporting to provide the unique probability function appropriate in any given epistemic state; starting in an imaginary primeval state of ignorance, one could, it was thought, describe with exactitude the gradual evolution in belief of the human mind. Recognizing clearly that this goal could not be achieved, Ramsey did not realize that one could nevertheless save the appearances by exhibiting (in appropriate settings) the purely qualitative aspects of inductive inference.

The key point is that previous attempts to explain induction had attempted to model the process by a unique description of prior beliefs (Bayes, 1764; Laplace, 1814; Broad, 1918; Johnson, 1924), or by a very narrow range of possibilities (Wrinch and Jeffreys, 1919; Johnson, 1932). De Finetti realized that because probability is a logic of consistency, one can never – *at a given instant of time* – uniquely dictate the partial beliefs of an individual; at most one can demand consistency. The essence of inductive behavior, in contrast, lies not in the specific beliefs that an individual entertains at any given point in time, but the manner in which those beliefs evolve over time. Let us pause briefly over this point.

I change my mind slowly; you do so with rapidity; you think I am pig-headed, I think you are rash. But neither of us is of necessity irrational. Disagreement is possible even if we share the same information; we may simply be viewing it in a different light. This is what happens every time the members of a jury disagree on a verdict. Of course it can be argued that the members of the jury do not share the same body of facts: each brings to the trial the sum total of his life experiences, and one juror tries to persuade another in part by drawing upon those experiences and thus enlarging the background information of their fellow jurors. It is the credibilist view of probability that if you knew what I knew, and I knew what you knew, then you and I would – or at least should – agree.

Such metaphysical stance may well be, as I. J. Good (1965, p. 7) says, "mentally health". But it is an article of faith of no real practical importance. None of us can fully grasp the totality of our own past history, experience, and information, let alone anyone else's. The goal is impossible; our information cannot be so encapsulated. But we *would* regard a person as irrational if we could not convert him to our viewpoint, no matter *how much* evidence he was provided with. From this perspective, irrationality is the *persistence* in a viewpoint in the face of mounting and cumulative evidence to the contrary.

The position that Ramsey adopts instead in his essay is a pragmatic one: Hume's demonstration ("so far as it goes") seems final, Keynes's view (that induction is a form of probable inference) cannot be maintained, but the result is hardly "a scandal to philosophy".[26] The Gordian knot is easily undone: induction is nothing other than "a useful habit, and so to adopt it is reasonable". (*This* was the part of Ramsey's position from which Keynes dissented.) Here Ramsey was merely restating Peirce's views; and the fact (as Ramsey freely acknowledges in a footnote) that the last five pages of his essay are "almost entirely based" on Peirce's writings suggests that they are not entirely the fruit of mature reflection.

The section on inductive inference in Ramsey's essay in fact clearly represents only a draft version of his initial thoughts on the subject: both the internal evidence of revision and the external evidence of further unpublished manuscripts make it evident that Ramsey continued to puzzle over the subject. Particularly tantalizing is a brief note in Ramsey's *Nachlass*, on the value of knowledge. Locke once wrote in the *Essay Concerning Human Understanding* that the person who "judges without informing himself to the utmost that he is capable, cannot acquit himself of *judging amiss*" (Book 2, Chapter 21, Section 67). Like the riddle of induction itself, this appears to be a natural intuition having an apparently elusive justification. But Ramsey was in fact able to provide such a justification, in terms of his theory of expected utility, thereby anticipating by several decades an argument later independently discovered by L. J. Savage (1954) and I. J. Good (1967).[27]

Also tantalizing is a fragment dealing with the rule of succession, deriving the predictive probability for a success on the next trial, given *n* previous successes, assuming only the *exchangeability* of the underlying sequence (Galavotti, 1992, pp. 279–281).[28] Here Johnson's influence is clearly evident. But lacking the de Finetti representation theorem, Ramsey was only able to show that this probability tends to one in the classical case considered by Bayes; and in order to establish more general results Ramsey had to leave the realm of exchangeability and return to the classical Laplacean world of prior distributions over chances.

Ramsey's failure to come to grips with induction – and de Finetti's failure to publish his work in English – had important consequences for philosophy. Keynes's treatment of the subject remained the standard (at least in English) for nearly three decades (until Carnap, 1950 and 1952); and most discussions of the relationship between probability and induction were dismissive if not openly contemptuous of the subjectivist viewpoint. Consider the state of affairs immediately after 1945: the books by Williams (1947), Russell (1948), and Kneale (1949) are ably written, often penetrating, but they address for the most part difficulties that the work of Ramsey and de Finetti had largely resolved, and the view these postwar books present of the interrelationship between probability and induction was completely obsolete at the time they were written. (Russell's book, for example, is an excellent summary of the state of the subject – as it existed in 1921 after the publication of Keynes's *Treatise*.) It was only when L. J. Savage arrived on the scene, and championed the work of Ramsey and de Finetti that the work of these two pioneers in subjective probability first received serious philosophical attention.[29]

NOTES

1. Persons achieving honors in the mathematical tripos at Cambridge were termed "Wranglers"; the "Senior Wrangler" was the top-ranking man of his year. (Man. After women were permitted to compete in 1882, they were not included in the ranking, but their position relative to it was given; e.g., "between 10 and 11". This could lead to absurdities, as in 1890, when P. G. Fawcett, a cousin of Littlewood, was listed as "above the Senior Wrangler"; see Littlewood, 1988, p. 134.) For an interesting statistical analysis of the examination scores for two years, see Galton (1869, pp. 14–18). J. E. Littlewood's *A Mathematician's Miscellany* gives an interesting personal view of the Cambridge examination system; see Littlewood (1988, pp. 3–5, 44–46, 83–87).
2. Galton described Ellis as a "brilliant senior wrangler . . . whose name is familiar to generations of Cambridge men as a prodigy of universal genius" (Galton, 1869, p. 18). There is a substantial body of manuscript material in the Wren Library, Trinity College, Cambridge, pertaining to Ellis (located, for the most part, in the Whewell collection). These include diaries, letters, and a number of mathematical and other notebooks.

3. For discussion of Ellis's theory, see Salmon (1980), Porter (1986, pp. 78–81), and Krüger (1987, pp. 68–70). In the 19th century Ellis was better known for his contributions to the theory of least squares, in particular Ellis (1844b), which extended Laplace's treatment of the subject and is the source of Todhunter's later account (1865, pp. 578–583). The paper is of interest from a modern perspective because of its discussion of the consistency of M-estimators; see Stigler (1973, p. 877–878), Plackett (1989, pp. 169–170). Ellis (1844d), not mentioned by Keynes, discusses Lagrange's solution to the problem of the duration of play; see Hald (1990, Chapter 23). For discussion of Ellis (1850), see Boole (1857, pp. 350–353).
4. One of Forbes's correspondents, however, also espoused the frequentist position. Sir George Airy, commenting on a draft of Forbes's reply to Herschel, remarked on a passage: "This is the only place in which you have adverted to the long run of many trials as entering into a chance-problem: and I think the want of more energetic reference to it is a defect in your paper. I think that that consideration is the foundation of all calculations of probabilities" (Shairp, et al. 1873, p. 476).
5. This may have been in part because Ellis wrote primarily as a critic of the Laplacean position, and did not attempt to provide – as Venn did later – a detailed alternative. Ellis's frequentist views appear to represent a second stage in his thinking, arising from an earlier dissatisfaction with the Laplacean attempt to quantify belief. His diary reveals that he had begun to work on the subject as early as April 1840; knowing of Whewell's forthcoming book on induction, he wrote to the Master of Trinity on 8 April:

I have been thinking of putting into the form of a little essay, some ideas on the application to natural philosophy of the doctrine of probabilities. I should attempt to point out the impossibility of a strict numerical estimate of the force of belief or interpretation, and at the same time make some remarks on the boundary between subjects capable of being treated mathematically and those which are not so. I should next consider the nature of the fundamental reasoning of the doctrine of probabilities a posteriori, and endeavour to show the vicious circle which it appears to involve, and then passing to the analogy on which Condorcet seems to rely, should try to examine its accuracy and authority. Lastly I should make some remarks, on the vague and arbitrary way, in which phenomena more or less similar are classed together as identical.

Ellis noted Whewell's response in his diary (entry for 9 April):

I wrote to Whewell, stating my notion of writing a little essay on probabilities, and asking if his work on the Philosophy of induction would interfere with it. I had a very civil answer today – saying he was glad to hear of my intention and wished me to persevere, as "he was sure I would throw light on it" ... Whewell's letter showed that one of the most arrogant of men of science was ready to acknowledge one as a fellow labourer in a favorite field of speculation ...

(The above quotations appear by the kind permission of the Master and Fellows of Trinity College Cambridge.)

6. The quotation is from the entry in the British *Dictionary of National Biography*, 1921–1930, pp. 869–870. The reference is to Henry Thomas Buckle (1821–1862), the author of *History of Civilisation in England* (vol. I, 1857; vol. 2, 1861); the entry on Buckle in the *Encyclopedia of Philosophy* provides a useful summary of his work. For the debate provoked by Buckle's discussion of statistical regularity in human affairs, see Porter (1986, pp. 60–70, 164–177), Stigler (1986, pp. 226–228), Krüger (1987, pp. 76–78), Hacking (1990, pp. 123–132).

7. First edition reviewed by C. S. Peirce, *North American Review* 105 (1867), p. 317. Second edition reviewed by T. V. Charpentier, *Revue philosophique 6* (1878) ("La logique du hasard d'apres M. John Venn"). Third edition reviewed by W. E. Johnson, *Mind* 49 (1888), pp. 268–280.

8. Fisher (1956, Chapter 2); see Zabell (1989b, pp. 250–251) for discussion.

9. The two papers Venn (1889 and 1891) bear witness to Venn's increasing interest in empirical statistical studies at this time.

10. It may provide a measure of consolation to some to know that in Johnson's time, just as in our own, advancement could sometimes be difficult, even for persons of unquestioned ability, highly regarded by their peers. On May 8, 1888, John Venn wrote to Francis Galton of the difficulty of finding suitable employment for Johnson despite the high opinion that Venn held of him.

   (From a letter in the Galton Papers at University College London. I thank Stephen Stigler for drawing my attention to the letter.)

11. The Apostles (technically, the Cambridge Conversazione Society) was (and is) a semi-secret discussion society at Cambridge, founded in 1820 (see generally Levy, 1979). It was distinguished from other such societies by the extraordinarily high intellectual caliber of membership. (For example, among the 46 members elected between May 1884 and the outbreak of war in 1914 were Whitehead, McTaggart, Russell, Moore, Roger Fry, G. M. Trevelyan, G. H. Hardy, E. M. Forster, Lytton Strachey, Leonard Woolf, Rupert Brooke, and Ludwig Wittgenstein.) In some cases prospective members ("embryos") might be identified even before they came to Cambridge (for example, by headmasters who had been members during their own undergraduate days at Cambridge). Membership was for life, although attendance at Saturday meetings ceased being compulsory after becoming an "angel". Ramsey was, not surprisingly, a member (as was Braithwaite), and his "Epilogue" paper (Braithwaite, 1931, pp. 281–292) was originally read at one of their meetings.

    One of the less happy members of the society in later years was Sir Anthony Blunt; in the wake of his unmasking as the so-called "fourth man" in the Philby spy network, knowledge of and information about the society became widespread. (But its secrecy has at times been somewhat exaggerated: Harrod, 1951, pp. 69–75 and *passim*, for example, describes it in some detail.)

12. Read perhaps 23 January 1904; see Skidelsky (1983, pp. 152–154). The exact date of the paper is controversial; Moggridge (1992, Chapter 5) argues on the basis of internal evidence that a later date (1906 or 1907) is more probable. For the influence of Moore's *Principia Ethica* on Keynes, see Harrod (1951, pp. 75–81), Levy (1979, pp. 239–246), Skidelsky (1983, Chapter 6). The ethical implications of probability are discussed by Keynes in Chapter 26 of the *Treatise*.

In 1938 Keynes read a paper entitled "My Early Beliefs" to a private gathering of friends (the so-called "Bloomsbury Memoir Club"); it appeared in print only after his death (Keynes, 1949). The paper, which describes the influence of Moore on Bloomsbury, has evoked strong reactions and no small amount of controversy; see Skidelsky (1983, Chapter 6), Moggeridge (1992, Chapter 5) for an analysis and discussion of further literature.

13. In a letter to Alfred Marshall in 1910, Keynes wrote that work on the *Treatise* had "occupied all my spare time for the last 4 years" (Harrod, 1951, p. 133); see generally Harrod (1951, pp. 127–8, 132–3), Skidelsky (1983, p. 233).

14. For the dispute with Pearson, see Harrod (1951, pp. 154–155), Skidelsky (1983, pp. 223–227).

15. Harrod (1951, p. 304).

16. Shortly after his dissertation was initially rejected, Keynes wrote to G. L. Strachey: "Johnson's report is almost as favorable as it could possibly be. I spent most of Sunday talking to him, and he had made a great number of very important criticisms, which, with the exception of one fundamental point, are probably right, and practically presented me with the fruits of his own work on the subject which have extended over years. On the pure logic of it he is, I think, quite superb and immensely beyond anyone else" (Harrod, 1951, p. 126).

17. Reviewed by C. D. Broad, *Mind* (N. S.) 31 (1922), pp. 72–85; H. Jeffreys, *Nature* 109 (1922), 132–3; Bertrand Russell, *Mathematical Gazette* 9 (1922); C. P. Sanger, *New Statesman*, 17 September 1921; Émile Borel, *Revue philosophique* (1924).

18. The letter has not, to my knowledge, been previously published; it appears here by the kind permission of the Provost and Fellows of King's College. (Unpublished writings of J. M. Keynes copyright The Provost and Scholars of King's College, Cambridge 1993.) Apart from its intrinsic historical interest, it strikingly illustrates the (here self-conscious) philosophical continuity at Cambridge in the foundations of probability that is one of the themes of the present essay.

19. For discussion of Keynes's theory, see Harrod (1951, pp. 133–141, 651–6), O'Donnell (1989), Moggeridge (1992, Chapter 6).

20. The circumstances surrounding the expulsion are discussed in detail by Hardy (1942); see also Levy (1979, pp. 287–288).

21. For Ramsey's relations with Wittgenstein, see Monk (1990). For details concerning Ramsey's translation of the *Tractatus*, see Monk (1990, p. 205).

22. As Knobloch (1987, p. 221) notes, the idea of interpreting degrees of belief in behavioral terms can be found two years earlier, in Borel's review of Keynes's *Treatise* (Borel, 1924). Note however that Ramsey himself makes no claim to originality; indeed, his language ("the old established way of measuring a person's belief") makes it clear that he does not regard the basic idea as in any way novel.

23. In his *Treatise on Probability* (Chapter 26), Keynes had criticized "the doctrine that the 'mathematical expectations' of alternative courses of action are the proper measures of our degrees of preference" (Keynes, 1921, p. 321). Ramsey's clever approach short-circuited Keynes's entire critique by turning matters on their head: mathematical expectations are no longer the "measures of our degrees of preference"; preferences become rather the instrument by which such expectations are determined.

24. For further discussion of Ramsey's method, and a critical analysis of the concept of the ethically neutral proposition, see Jeffrey (1983, pp. 46–52, 55–57); see also Sneed (1966).
25. In a recent revisionist attempt, O'Donnell (1989, pp. 139–148) has argued that this was not in fact the case, but his argument, given the clear language employed by Keynes, is not convincing.
26. This last phrase was a (perhaps derisive) allusion to a lecture given by C. D. Broad a month earlier (on October 5th), during the Bacon tercentenary celebration held in the Cambridge Senate House. Broad had then expressed the hope that the time would soon come when "Inductive Reasoning, which has long been the glory of Science, will have ceased to be the scandal of Philosophy" (see Hacking, 1980, p. 142).
27. Galavotti (1992, pp. 285–287); see Skyrms (1990, Chapter 4).
28. Thus, if the prior probability of k successes out of n + 1 in an exchangeable sequence is $\varphi(k)$, then the probability of observing n + 1 successes out of n + 1, given that the first n observed are all successes, is

$$\frac{n+1}{n+1+\frac{\varphi(n)}{\varphi(n+1)}}$$

(The general formula for a success on the (n + 1)-st trial in an exchangeable sequence, given k prior successes out of n, was published by de Finetti in 1937; see Zabell, 1989a, p. 305.)
29. I thank David Stove for his comments on an earlier draft of this paper; Anthony Edwards for providing a copy of the letter from Keynes to Venn quoted earlier; the Master and Fellows of Gonville and Caius College for their hospitality during a visit to Cambridge to study original manuscript material pertaining to the subject of this paper; Dr. Ronald Milne and the staff of the Wren Library of Trinity College Cambridge for their exemplary assistance when I consulted their collection; and Maria Carla Galavotti for her invitation to speak at the 1990 Ramsey conference at the University of Bologna (of which this paper is a direct consequence).

REFERENCES

Annan, N. G. (1955). The intellectual aristocracy. In J. H. Plumb (ed.). *Studies in Social History: A Tribute to G. M. Trevelyan*, London: Longmans and Green, pp. 241–287.
Bayes, T. (1764). An essay towards solving a problem in the doctrine of chances. *Philosophical Transactions of the Royal Society of London* **53**, 370–418.
Boole, B. (1857). On the application of the theory of probabilities to the Question of the combination of testimonies or judgments. *Transactions of the Royal Society of Edinburgh* 21. Reprinted in *Collected Logical Works*, vol. 1: *Studies in Logic and Probability*, 308–385. [References are to this edition.]
Borel, É. (1924). A propos d'un traité de probabilités. *Revue philosophique* 98, 321–326. English translation 1964, H. E. Kyburg and H. E. Smokler (eds.), *Studies in Subjective Probability*, New York: Wiley, pp. 45–60.
Braithwaite, R. B., ed. (1931). *The Foundations of Mathematics and Other Logical Essays*. London: Routledge and Kegan Paul.

Broad, C. D. (1918). The relation between induction and probability I. *Mind* 27, 389–404.

Broad, C. D. (1920). The relation between induction and probability II. *Mind* 29, 11–45.

Carnap, R. (1950). *Logical Foundations of Probability*. Chicago: University of Chicago Press.

Carnap, R. (1952). *The Continuum of Inductive Methods*. Chicago: University of Chicago Press.

Dale, A. I. (1991). *A History of Inverse Probability from Thomas Bayes to Karl Pearson*. New York: Springer-Verlag.

De Finetti, B. (1972). *Probability, Induction, and Statistics*, New York: Wiley.

De Morgan, A. (1847). *Formal Logic: Or the Calculus of Inference Necessary and Probable*. London: Taylor and Watton.

Edgeworth, F. Y. (1884). The philosophy of chance. *Mind* 9, 222–235.

Edgeworth, F. Y. (1922). The philosophy of chance. *Mind* 31, 257–283.

Ellis, Robert Leslie (1844a). On the foundations of the theory of probabilities. *Transactions of the Cambridge Philosophical Society* 8, Part 1, 1–6. (Read 14 February 1842.)

Ellis, Robert Leslie (1844b). On the method of least squares. *Transactions of the Cambridge Philosophical Society* 8, Part 1, 204–219. (Read 4 March 1844.)

Ellis, Robert Leslie (1844c). On a question in the theory of probabilities. *Cambridge Mathematical Journal* 4: 127.

Ellis, Robert Leslie (1844d). On the solutions of equations in finite differences. *Cambridge Mathematical Journal* 4, 182–190.

Ellis, Robert Leslie (1850). On an alleged proof of the "method of least squares". *Philosophical Magazine* 37.

Ellis, Robert Leslie (1854). Remarks on the fundamental principle of the theory of probabilities. *Transactions of the Cambridge Philosophical Society* 9, 605–607.

Fisher, R. A. (1922). On the mathematical foundations of theoretical statistics. *Phil. Trans. Roy. Soc. London A* 222, 309–368.

Fisher, R. A. (1925). Theory of statistical estimation. *Proceedings of the Cambridge Philosophical Society* 22, 700–725.

Fisher, R. A. (1956). *Statistical Methods and Scientific Inference*. New York: Hafner (2nd ed., 1959, 3rd ed., 1973).

Galavotti, Maria Carla, ed. (1992). *Frank Plumpton Ramsey: Notes on Philosophy, Probability and Mathematics*. Napoli: Bibliopolis.

Galton, F. (1869). *Hereditary Genius*. 2nd. ed. 1892; reprinted 1925, London: Macmillan. (Page references are to this edition.)

Good, I. J. (1965). *The Estimation of Probabilities: An Essay on Modern Bayesian Methods*. Cambridge, MA: M.I.T. Press.

Good, I. J. (1967). On the principle of total evidence. *British Journal for the Philosophy of Science* 17, 319–321.

Hacking, I. (1967). Slightly more realistic personal probability. *Philosophy of Science* 34, 311–325.

Hacking, I. (1980). The theory of probable inference: Neyman, Peirce and Braithwaite. In D. H. Mellor (ed.), *Science, Belief and Behaviour*, Cambridge, UK: Cambridge University Press, pp. 141–160.

Hacking, I. (1990). *The Taming of Chance*. Cambridge, UK: Cambridge University Press.

Hald, A. (1990). *A History of Probability and Statistics and Their Applications before 1750*. New York: Wiley.

Hardy, G. H. (1942). *Bertrand Russell and Trinity*. Cambridge, UK: Cambridge University Press.

Harrod, R. F. (1951). *The Life of John Maynard Keynes*. London: Macmillan.

Jeffrey, Richard C. (1968). Probable knowledge. In *The Problem of Inductive Logic* (I. Lakatos, ed.), Amsterdam: North-Holland, pp. 166–180.

Jeffrey, R. C. (1983). *The Logic of Decision*, 2nd ed. Chicago: University of Chicago Press.

Jeffreys, Sir Harold (1973). *Scientific Inference*, 3rd ed. Cambridge, UK: Cambridge University Press (1st ed. 1931, 2nd ed. 1957).

Johnson, W. E. (1924). *Logic, Part III. The Logical Foundations of Science*. Cambridge, UK: Cambridge University Press. [For a critical review, see C. D. Broad, "Mr. Johnson on the Logical Foundations of Science", *Mind* 33 (1924), 242–269, 367–384.]

Johnson, W. E. (1932). "Probability" *Mind* 41: 1–16 ("The Relations of Proposal to Supposal"), 281–296 ("Axioms"), 409–423 ("The Deductive and Inductive Problems").

Kamlah, A. (1987). The decline of the Laplacian theory of probability: a study of Stumpf, von Kries, and Meinong. In Krüger, L., Daston, L. J., and Heidelberger, M. (eds.), *The Probabilistic Revolution*, vol. 1: *Ideas in History*. Cambridge, MA: MIT Press, pp. 91–116.

Keynes, J. M. (1921). *A Treatise on Probability*. London: Macmillan.

Keynes, J. M. (1933). *Essays in Biography*. Revised edition 1951, edited by G. Keynes, Horizon Press. Reprinted 1963, New York: Norton.

Keynes, J. M. (1949). *Two Memoirs, Dr. Melchior: A Defeated Enemy and My Early Beliefs*. London: R. Hart-Davies.

Kneale, W. (1949). *Probability and Induction*. Oxford, UK: Clarendon Press.

Knobloch, E. (1987). Emile Borel as probabilist. In Krüger, L., Daston, L. J., and Heidelberger, M. (eds.), *The Probabilistic Revolution*, vol. 1: *Ideas in History*. Cambridge, MA: MIT Press, pp. 215–233.

Krüger, L. (1987). The slow rise of probabilism: philosophical arguments in the nineteenth century. In Krüger, L., Daston, L. J., and Heidelberger, M. (eds.), *The Probabilistic Revolution*, vol. 1: *Ideas in History*. Cambridge, MA: MIT Press, pp. 59–89.

Laplace, P. S. Marquis de (1814). *Essai philosophique sur les probabilités*. Paris: Courcier.

Levy, P. (1979). *Moore: G. E. Moore and the Cambridge Apostles*. New York: Oxford University Press.

Littlewood, J. E. (1953). *A Mathematician's Miscellany*. Cambridge University Press. Reprinted 1988, B. Bollobás (ed.), *Littlewood's Miscellany*, Cambridge, UK: Cambridge University Press. (References are to this edition.)

Moggridge, D. E. (1992). *Maynard Keynes: An Economist's Biography*. London and New York: Routledge.

Monk, R. (1990). *Ludwig Wittgenstein: The Duty of Genius*. New York: The Free Press.

O'Donnell, R. M. (1989). *Keynes: Philosophy, Economics, and Politics: The Philosophical Foundations of Keynes's Thought and Their Influence on His Economics and Politics*. London: Macmillan.

Pearson, K. (1892). *The Grammar of Science*. London: Walter Scott.

Plackett, R. L. (1989). The influence of Laplace and Gauss in Britain. *I. S. I. –47th Session*, Paris, August 29-September 6, 1989, pp. 163–176.

Porter, T. M. (1986). *The Rise of Statistical Thinking 1820–1900*. Princeton, NJ: Princeton University Press.

Ramsey, F. P. (1922). Mr Keynes on probability. *The Cambridge Magazine* 11, 3–5; reprinted 1989, *Brit. J. Phil. Sci.* 40, 219–222.

Russell, B. (1948). *Human Knowledge: Its Scope and Limits*. New York: Simon and Schuster.

Sahlin, N.-E. (1990). *The Philosophy of F. P. Ramsey*. Cambridge, UK: Cambridge University Press.

Salmon, Wesley C. (1980). Robert Leslie Ellis and the frequency theory. In *Pisa Conference Proceedings*, vol. 2 (J. Hintikka, D. Gruender, and E. Agazzi, eds.), Dordrecht: D. Riedel, pp. 139–143.

Savage, L. J. (1954). *The Foundations of Statistics*. New York: Wiley.

Shairp, J. C., Tait, P. G., and Adams-Reilly, A. (1873). *Life and Letters of James David Forbes, F. R. S.* London.

Skidelsky, R. (1983). *John Maynard Keynes: A Biography*. London: Macmillan.

Skyrms, Brian (1990). *The Dynamics of Rational Deliberation*. Cambridge, MA: Harvard University Press.

Sneed, J. (1966). Strategy and the logic of decision. *Synthese* **16**, 270–283.

Stigler, Stephen M. (1973). Simon Newcomb, Percy Daniell, and the history of robust estimation 1885–1920. *J. American Statistical Association* 68, 872–879.

Stigler, Stephen M. (1986). *The History of Statistics*. Cambridge, MA: Harvard University Press.

Todhunter, Isaac (1865). *A History of the Mathematical Theory of Probability from the Time of Pascal to that of Laplace*. London: Macmillan. Reprinted 1949, 1965, New York: Chelsea.

Venn, J. (1866). *The Logic of Chance*. London: Macmillan (2nd ed., 1876; 3rd ed., 1888).

Venn, John (1889). Cambridge anthropometry. *Journal of the Anthropological Institute* 18, 140–54.

Venn, John (1891). On the nature and uses of averages. *Journal of the Royal Statistical Society* 54, 429–48.

Walton, W., ed. (1863). *The Mathematical and Other Writings of Robert Leslie Ellis, M. A.* Cambridge, UK: Deighton, Bell.

Williams, D. C. (1947). *The Grounds of Induction*. Cambridge, MA: Harvard University Press.

Wrinch, D. M. and Jeffreys, H. (1919). On certain aspects of the theory of probability. *Philosophical Magazine* (Series 6) 38, 715–731.

Wrinch, D. M. and Jeffreys, H. (1921). On certain fundamental principles of scientific inquiry. *Philosophical Magazine* (Series 6) 42, 369–390; 45, 368–374.

Zabell, S. L. (1982). W. E. Johnson's 'sufficientness' postulate. *Annals of Statistics* 10, 1091–1099.

Zabell, S. L. (1988). Symmetry and its discontents. In B. Skyrms and W. L. Harper (eds.), *Causation, Chance, Credence*, vol. 1, Dordrecht: Kluwer, pp. 155–190.

Zabell, S. L. (1989a). The rule of succession. *Erkenntnis* 31, 283–321.

Zabell, S. L. (1989b). R. A. Fisher on the history of inverse probability. *Statistical Science* 4, 247–263.

7

R. A. Fisher on the History
of Inverse Probability

Abstract. R. A. Fisher's account of the decline of inverse probability methods during the latter half of the nineteenth century identifies Boole, Venn and Chrystal as the key figures in this change. Careful examination of these and other writings of the period, however, reveals a different and much more complex picture. Contrary to Fisher's account, inverse methods – at least in modified form – remained theoretically respectable until the 1920s, when the work of Fisher and then Neyman caused their eclipse for the next quarter century.

Key words and phrases: R. A. Fisher, inverse probability, history of statistics.

R. A. Fisher was a lifelong critic of inverse probability. In the second chapter of his last book, *Statistical Methods and Scientific Inference* (1956), Fisher traced the history of what he saw as the increasing disaffection with Bayesian methods that arose during the second half of the nineteenth century. Fisher's account is one of the few that covers this neglected period in the history of probability, in effect taking up where Todhunter (1865) left off, and has often been cited (e.g., Passmore, 1968, page 550, n. 7 and page 551, n. 15; de Finetti, 1972, page 159; Shafer, 1976, page 25). The picture portrayed is one of gradual progress, the logical lacunae and misconceptions of the inverse methods being steadily recognized and eventually discredited.

But on reflection Fisher's portrait does not appear entirely plausible. Edgeworth and Pearson, two of the most distinguished statisticians of the generation immediately prior to Fisher's, were both sympathetic to inverse methods; and indeed, as will be discussed later, Bayesian methods were widely taught and employed in England and elsewhere until the 1930s. It was only then that Fisher and Neyman simultaneously administered a nearly lethal blow to Bayesian statistics, one from which it was not to recover until

Reprinted with permission from *Statistical Science* 4, no. 3 (1989): 247–263.

the publication, nearly a quarter of a century later, of Savage's *Foundations of Statistics* in 1954.

How was such a disparity between Fisher's account and historical reality possible? Careful examination of Fisher's own evidence for his claims reveals an interesting story, telling us perhaps in some ways as much about Fisher as it does about the period he discusses.

1. FISHER'S ACCOUNT

Fisher cites three major authorities for the decline in the prestige of inverse methods: Boole, Venn and Chrystal. He had done so repeatedly in earlier papers (Fisher, 1922, pages 311 and 326; 1930, page 531; 1936a, page 248; 1951, page 49), and his account in *Statistical Methods and Scientific Inference* (*SMSI*) is an elaboration on these earlier, fragmentary comments. The following passages give the flavor of his argument:

The first serious criticism was developed by Boole in his "Laws of Thought" in 1854.... Boole's criticism worked its effect only slowly. In the latter half of the nineteenth century the theory of inverse probability was rejected more decisively by Venn and by Chrystal.... [Fisher, 1936a, page 248]

[Venn's criticisms of the Rule of Succession], from a writer of his weight and dignity, had an undoubted effect in shaking the confidence of mathematicians in its mathematical foundation. [*SMSI*, page 25]

Perhaps the most important result of Venn's criticism was the departure made by Professor G. Chrystal in eliminating from his celebrated textbook of *Algebra* the whole of the traditional material usually presented under the headings of *Inverse Probability* and of the *Theory of Evidence*. [*SMSI*, page 29]

Fisher did not try to overstate the immediate impact of these criticisms. He noted "the slowness with which the opinions of Boole, Venn and Chrystal were appreciated," and drew attention to the defenses of inverse probability mounted by Edgeworth (1908) and Pearson (1920). Fisher was not always consistent on this point, however. Writing a few years later, he describes the supposed rejection of inverse probability in England as occurring "abruptly and dramatically" (Fisher, 1958, page 273), and uses the phrase "as late as 1908" in referring to Edgeworth's paper. Nevertheless, Fisher's earlier reference to the "decisive criticisms to which [the methods of inverse probability] had been exposed at the hands of Boole, Venn, and Chrystal" (1922, page 326), and his assertion that "[t]hese criticisms appear to be unanswerable, and the theory of inverse probability . . . is now almost universally abandoned" (1951, page 49) capture the basic points of his more extended account in *SMSI*: these were the key critics, their criticisms were well-founded and they were largely responsible for the decline and fall of inverse probability.

The reader, however, who turns to Boole, Venn and Chrystal to see what they actually wrote – how accurately Fisher represents their views and to what extent they actually support Fisher's position – will find the result surprising.

2. BOOLE

Boole, Fisher says, was the first to seriously criticize "Bayes' doctrine" (Fisher, 1936a, page 249; cf. Fisher, 1951, page 49). This was only partially true. Robert Leslie Ellis had a decade earlier formulated a frequentist theory of probability (Ellis, 1844) and criticized the Laplacian approach to inference on a number of grounds including *ex nihilo nihil* (out of nothing, nothing) – i.e., no inference at all is warranted in a situation of complete ignorance. John Stuart Mill had also been, albeit briefly, a critic. In addition, both Jakob Friedrich Fries in Germany and Antoine Augustin Cournot in France had earlier discussed objective or frequentist theories of probability and attacked uncritical applications of inverse probability. (Cournot was less strident than an earlier French tradition represented by Destutt de Tracy, Poinsot and Auguste Comte; see generally Porter (1986, pages 77–88) and Stigler (1986, pages 194–200). Fisher sometimes appears to have been surprisingly unfamiliar with nineteenth century developments outside of England, and this often gives his historical discussions a somewhat insular flavor. Thus, he also makes no mention of Bertrand, although Bertrand's *Calcul des probabilités* (1st edition, 1889) sharply criticized inverse methods and was without question highly influential.)

Boole's criticisms were a natural outgrowth of his philosophical view that probability is a logical relation between propositions. In this he was very close to De Morgan; both De Morgan's *Formal Logic* (1847) and Boole's *Investigation of the Laws of Thought* (1854) treated probability as a branch of logic. But while De Morgan and others believed that any event possessed – at least in principle – a definite numerical probability relative to a given body of information (e.g., De Morgan, 1847, page 178; Donkin, 1851, pages 354–355), Boole argued that, lacking sufficient information, the probabilities of some events were indeterminate.

This was an important point, because a major defense of uniform priors in Boole's day was a challenge to doubters to produce a more plausible alternative: "A person who should dispute the propriety of dividing our belief equally amongst hypotheses about which we are equally ignorant, ought to be

144

refuted by asking him to state which is to be preferred. He must either admit the proposed law, or maintain that there is no law at all" (Donkin, 1851, page 355). The latter is precisely what Boole did. As a result, he was able to criticize previous treatments which attempted to sidestep indeterminacy by hypothesis:

It has been said, that the principle involved in the above and in similar applications is that of the equal distribution of our knowledge, or rather of our ignorance – the assigning to different states of things of which we know nothing, and upon the very ground that we know nothing, equal degrees of probability. I apprehend, however, that this is an arbitrary method of procedure. [Boole, 1854, page 370]

Boole supported this criticism by making the simple but telling point that in some cases the principle could be applied in more than one way to the same problem, resulting in two or more conflicting probability assignments. For example, Bayes had argued that "in the case of an event concerning the probability of which we absolutely know nothing antecedently to any trials made concerning it . . . I have no reason to think that, in a certain number of trials, it should rather happen any one possible number of times than another;" i.e., that

$$P[S_n = k] = 1/(n + 1), \quad k = 0, 1, \ldots, n$$

(where S_n denotes the number of successes in n trials). But, as Boole pointed out, one could equally well argue that all *sequences* of outcomes in n trials should be viewed as equally likely, resulting in an entirely different probability assignment. (Bertrand's paradox (involving random choice of a chord) made the same point for a continuous variate (Bertrand, 1907, pages 4–5). Along the same lines, Fisher was fond of pointing out that uniform priors on continuous parameter spaces were not invariant under all continuous transformations (e.g., Fisher, 1956, page 16).)

This was an important observation, but it did not compel abandonment of the principle of indifference. It did provide a warning that naive application of the principle could lead to paradoxes and inconsistencies, and during the next century many philosophers – notably von Kries, Keynes, Jeffreys, and Carnap – undertook to refine it in an attempt to avoid them (von Kries, 1886; Keynes, 1921; Jeffreys, 1939; Carnap, 1950).

Nor did Boole himself advocate abandonment of the principle. This might not have been apparent to someone reading only *The Laws of Thought*, for there mention of the principle is indeed limited to a discussion of its improper usage. But Boole repeatedly returned to the foundations of probability in his subsequent papers, and Fisher would scarcely have found himself in agreement with Boole's later opinions.

In his last, perhaps most considered thoughts on the subject, Boole wrote that:

All the procedure of the theory of probabilities is founded on the mental construction of the problem from some hypothesis, either, first, of events known to be independent; or secondly, of events of the connexion of which we are totally ignorant; so that upon the ground of this ignorance, we can again construct a scheme of alternatives all equally probable, and distinguished merely as favouring or not favouring the event of which the probability is sought. In doing this we are not at liberty to proceed arbitrarily. We are subject, first, to the formal *Laws of Thought*, which determine the possible conceivable combinations; secondly, to that principle, more easily conceived than explained, which has been differently expressed as the "principle of non-sufficient reason," the "principle of the equal distribution of knowledge or ignorance," and the "principle of order." We do not know that the distribution of properties in the actual urn is the same as it is conceived to be in the ideal urn of free balls, but the hypothesis that it is so involves an equal distribution of our actual knowledge, and enables us to construct the problem from ultimate hypotheses which reduce it to a calculation of combinations. [Boole, 1862, pages 389–390 of 1952 edition]

Obviously Fisher could never have accepted this view of the nature of probability, or the imprimatur it bestows upon the use of the principle of insufficient reason. (In the third edition of *SMSI*, Fisher added a subsection on Todhunter, who had emphasized that "in Bayes's own problem, we *know* that *a priori* any position of *EF* between *AB* and *CD* is equally likely; or at least we know what amount of assumption is involved in this supposition. In the applications which have been made of Bayes's theorem, and of such results as that which we have taken from Laplace in Art. 551, there has however often been no adequate ground for such knowledge or assumption" (Todhunter, 1865, pages 299–300). Fisher praised Todhunter's emphasis on the necessity for a factual rather than an axiomatic basis for prior probabilities. Nevertheless, because of Todhunter's use of the qualifying phrase "or at least we know what amount of assumption is involved in this supposition," Fisher concluded that "Near as he came to clarifying the situation, Todhunter's name cannot properly be added to those who finally succeeded in extricating the mathematical thought of the mid-nineteenth century from its bewildering difficulties." This suggests that Fisher would have been highly critical of Boole's later remarks.)

Had Boole changed his mind? He claims not, for he added in a footnote:

... I take this opportunity of explaining a passage in the *Laws of Thought*, page 370, relating to certain applications of the principle. Valid objection lies not against the principle itself, but against its application through arbitrary hypotheses, coupled with the assumption that any result thus obtained is necessarily the true one. The application of the principle employed in the text and founded upon the general theorem of development in Logic, I hold to be *not* arbitrary.

The distinction that Boole intends pits the so-called principle of *insufficient* reason, against what was later described as the "principle of *cogent* reason," i.e., that the probabilities assigned to alternatives should be taken to be equal if the information about those alternatives equally favors each (as Boole puts it, if there is "an equal distribution of our *actual* knowledge"). In any case, it is clear that Boole was not an opponent of the use of some form of the principle, and was opposed instead to what he considered its uncritical application. (As Keynes (1921, page 167) and many others have noted, Boole's writings on probability are also marred by a systematic confusion between two different meanings of independence. Hailperin (1976) provides a helpful guide through the thicket.)

3. VENN

John Venn was a Cambridge logician, best known today for his popularization of "Venn diagrams," and in his own day for his influential textbook *Symbolic Logic* (1st edition, 1881; 2nd edition, 1894). Yet in terms of originality and long-term impact, Venn's most important work is his *Logic of Chance* (1st edition, 1866), which gave the first detailed discussion in English of a frequentist theory of probability, as well as a careful critique of the earlier Laplacean position, including both its use of uniform priors and the consequences that follow from such an assumption. It was Venn's discussion of one of these consequences that Fisher examined in *SMSI*.

3.1. The Rule of Succession

Laplace's "Rule of Succession" states (in brief) that an event which has occurred n times in succession will recur the next time with probability $(n + 1)/(n + 2)$. Venn ridiculed the Rule of Succession, pointing out a variety of cases where it contradicted common sense (rain on three successive days; death caused by administered strychnine on three separate occasions; people answering a false call of fire on three different occasions). While Fisher cited Venn with general approbation, he took issue with him on this particular point. As Fisher was quick to point out, "such a rule can be based on Bayes' theorem only on certain conditions." In particular, the successive trials must be *independent*, which is certainly not the case in two of Venn's examples.

Fisher was in fact highly critical of Venn: Venn "perhaps was not aware that it [the Rule of Succession] had a mathematical basis demonstrated by Laplace;" "there is no doubt that Venn in this chapter uses arguments of a quality which he would scarcely have employed had he regarded the matter as one open to rational debate;" Venn's examples "seem to be little more than rhetorical sallies intended to overwhelm an opponent with ridicule;" and "by his eagerness to dispose of [the Rule of Succession] . . . he became uncritical of the quality of the arguments he used."

In order to judge the validity and persuasiveness of Venn's treatment, in the light of Fisher's comments, it is natural to turn to Venn's original discussion, in order to read his arguments in context. The reader who turns to the reprinted edition of *The Logic of Chance*, however, will find to his surprise that although Venn does indeed devote an entire chapter to the Rule of Succession, the passages that Fisher quotes are nowhere to be found!

The solution to this puzzle, however, is not difficult. *The Logic of Chance* went through three editions – 1866, 1876, and 1888, the currently available Chelsea reprint being the last of these. Although Fisher does not indicate in *SMSI* which edition he consulted, a comparison of editions reveals that Fisher was quoting from the 2nd edition, a copy of which he may have owned (this edition is cited in Fisher, 1955).

This was not a minor matter, inasmuch as Venn made substantial revisions in both the second and third editions of *The Logic of Chance*. (A comparative study of the three editions of the *Logic*, tracing the evolution of Venn's thought, would be of considerable interest. Salmon (1981) discusses some differences, but largely confines his attention to the first edition.)

In this instance, between the second and third editions Venn made major changes in the chapter on the Rule of Succession, taking out precisely the examples that Fisher so vehemently objected to. It is natural to assume that between editions a colleague or correspondent – very likely Edgeworth, whose help is acknowledged in the preface – voiced criticisms very similar to Fisher's; indeed, Venn's revision addresses precisely the points raised by Fisher: the mathematical assumptions underlying the derivation of the rule, and their possible empirical validity.

Another puzzle is the tenor of Fisher's discussion. Fisher was in a certain sense very "political" in his writings; often quick to attack the opposition, he seldom expressed in print reservations he might express to close friends and allies. That he should sharply criticize an ally like Venn seems strangely inconsistent with his usual practice. In this case, however, a simple explanation suggests itself.

Venn's criticisms were not of the inverse rule per se, but its mathematical consequence, the Rule of Succession. Thus, the examples he adduces, to the extent that they discredit the Rule of Succession, also discredit *any* form of inference that gives rise to the Rule of Succession.

And that would include fiducial inference. For in the next chapter of *SMSI*, during a discussion of the application of the fiducial argument to discontinuous data, Fisher notes that:

An odd consequence of the analysis developed above is that the Rule of Succession derivable from the particular distribution of probability *a priori*

$$\frac{dp}{\pi\sqrt{pq}},$$

namely that the probability of success in the next trial is

$$\frac{a + \frac{1}{2}}{a + b + 1}$$

is justifiable, at least to a remarkably high approximation, in the absence of any knowledge *a priori*; and this although the corresponding complete distribution *a posteriori* is not so justifiable. [Fisher, 1956, page 68]

Thus an attack on the Rule of Succession was actually an indirect attack on the fiducial argument as well and, as such, had to be met. But Fisher was curiously coy about the matter. In his discussion of Venn, no mention is made of the fact that the Rule can be so justified, only that Venn's criticisms were specious. And when Fisher derives the Rule as an approximate consequence of the fiducial argument, no mention is made of Venn's criticisms.

There is no clear evidence whether Fisher was aware of the third edition of Venn's *Logic of Chance*. Certainly, had he seen it, he would have approved of the changes Venn made in the chapter on the Rule of Succession. But Venn made a number of other revisions as well, one of which Fisher would most certainly not have approved.

3.2. *Probability and Listerism*

In 1879, Dr. Donald MacAlister posed the following question in the pages of the *Educational Times*:

Of 10 cases treated by Lister's method, 7 did well and 3 suffered from blood-poisoning: of 14 treated with ordinary dressings, 9 did well and 5 had blood poisoning; what are the odds that the success of Lister's method was due to chance?

Due to the small sizes of the samples involved, the large-sample methods then available for analyzing such differences were inapplicable, and the Bayesian solution advocated by MacAlister involved assigning independent uniform priors to the two unknown binomial proportions (see generally Winsor, 1947).

In the third edition of the *Logic of Chance*, Venn included a discussion of MacAlister's question. Consistency required that Venn reject MacAlister's approach, yet Venn was obviously uncomfortable with a position that no inference could be drawn. The result was a surprising reversal. Venn describes the example as illustrating those cases which afforded "[t]he nearest approach to any practical justification for [inverse] judgments," and approves of MacAlister's treatment of it as a 'bag and balls' problem, being "the only reasonable way of treating the problem, if it is to be considered capable of numerical solution at all" (Venn, 1888, pages 186–187). Thus far Fisher might still have had no difficulty. But then Venn went on to add:

Of course the inevitable assumption has to be made here about the equal prevalence of the different possible kinds of bag – or, as the supporters of the justice of the calculation would put it, of the obligation to assume the equal *a priori* likelihood of each kind – but I think that in this particular example the arbitrariness of the assumption is less than usual. This is because the problem discusses simply a balance between two extremely similar cases, and there is a certain set-off against each other of the objectionable assumptions on each side. Had *one* set of experiments only been proposed, and had we been asked to evaluate the probability of continued repetition of them confirming their verdict, I should have felt all the scruples I have already mentioned. But here we have got two sets of experiments carried on under almost exactly similar circumstances, and there is therefore less arbitrariness in assuming that their unknown conditions are tolerably equally prevalent.

Venn's logic is difficult to follow; the last three sentences seem more a rationalization than a carefully thought-out argument. (This is hardly surprising, since the position Venn now takes is totally incompatible with the one he had previously adopted.) What *is* clear is that Fisher would have rejected it entirely. Todhunter had been excluded from the pantheon of clarification for defending Bayes's postulate when "we know what amount of assumption is involved in this supposition." Fisher's reaction to Venn's apostasy can only be conjectured.

4. CHRYSTAL

Chrystal, Fisher says, "does not discuss the objections to this material [inverse probability and the theory of evidence]." This was only partly true. Although

Chrystal did not elaborate in his *Algebra* on his reasons for omitting inverse probability, he did return to the subject five years later and present his objections in detail. It was easy to overlook this paper of Chrystal's, for it appeared in the *Transactions of the Actuarial Society of Edinburgh* (1891), a journal not widely available, as anyone who attempts to consult Chrystal's paper will readily find. In his 1891 paper, Chrystal spelled out his views on probability, views that Fisher would have found a serious embarrassment.

Fisher had always been at pains to emphasize that he had no objection to the use of Bayes's theorem, only to its unwarranted application in situations where information justifying the use of a prior was unavailable; in particular, Fisher objected to the principle of insufficient reason to assign priors (e.g., *SMSI*, page 20). Chrystal's objections, ironically, were exactly the opposite: he did not object to the use of ignorance priors, but thought that given a prior, Bayes's theorem could generate an incorrect answer! He writes:

Perhaps the following ... will make the absurdity of the supposed conclusion of the Inverse Rule still clearer.

A bag contains three balls, each of which is either white or black, all possible numbers of white being equally likely. Two at once are drawn at random and prove to be white; what is the chance that all the balls are white?

Any one who knows the definition of mathematical probability, and who considers this question apart from the Inverse Rule, will not hesitate for a moment to say that the chance is $\frac{1}{2}$; that is to say, that the third ball is just as likely to be white as black. For there are four possible constitutions of the bag:

	1°	2°	3°	4°
W	3	2	1	0
B	0	1	2	3

each of which, we are told, occurs equally often in the long run, and among those cases there are two (1° and 2°) in which there are two white balls, and among these the case in which there are three white occurs in the long-run just as often as the case in which there are only two.

Chrystal then goes on to correctly calculate that, in contrast, the "application of the Inverse Rules" leads to posterior odds of 3 to 1 in favor of the third ball being white, and concludes:

No one would say that if you simply put two white balls into a bag containing one of unknown colour, equally likely to be black or white, that this action raised the odds that the unknown ball is white from even to 3 to 1. It appears, however, from the

Inverse Rule that if we find out that the two white balls are in the bag, not by putting them in, but by taking them out, it makes all the difference.

Indeed it does. Chrystal's error is exactly the point of the closely related *Bertrand box paradox* (Bertrand, 1907, pages 2–3).

In the light of this fundamental misunderstanding, Chrystal's objections to inverse probability can scarcely be described as intellectually devastating. He was merely one of many (e.g., D'Alembert and Mill) whose intellectual attainments in other areas led him to uncritically accept his own untutored probabilistic intuitions. As Jevons once noted, "It is curious how often the most acute and powerful intellects have gone astray in the calculation of probabilities" (Jevons, 1877, page 213). (In 1893, shortly after Chrystal read his paper before the Actuarial Society of Edinburgh, John Govan read a paper before the same body, pointing out the errors and confusions in Chrystal's paper. It went unpublished, however, until 1920, when the eminent mathematician E. T. Whittaker read a similar exposé before the London Faculty of Actuaries (Whittaker, 1920).)

The conclusion to this episode in the history of the history of statistics is somewhat bizarre. Of his trinity of authorities – Boole, Venn and Chrystal – Fisher thought Boole was an opponent of inverse methods, but Boole was not; Venn was an opponent, but only in part; and Chrystal was an unqualified opponent, but on grounds Fisher would have found repugnant, had he known of them.

5. INVERSE PROBABILITY FROM 1880 TO 1930

What was the actual impact of these critics? Contrary to what Fisher suggests, they did not eliminate inverse methods. Edgeworth and Pearson, perhaps the two most prominent English statisticians of the generation immediately preceding Fisher's, both remained sympathetic to Bayesian methods. Moreover, we have the testimony of Fisher himself that he had "learned it at school as an integral part of the subject, and for some years saw no reason to question its validity" (Fisher, 1936a, page 248). Indeed, he had to "plead guilty in my original statement of the Method of Maximum Likelihood [Fisher, 1912] to having based my argument upon the principle of inverse probability ... " (Fisher, 1922, page 326).

The real effect of Boole, Venn, and Chrystal and other critics appears rather to have been to cause the exponents of inverse methods to hedge their claims for the theory. For example, William Allen Whitworth, the author of a popular nineteenth century textbook *Choice and Chance*, dealt with objections to the rule of succession by conceding that expressions such as "entirely unknown" in its formulation were "vague." He proposed that they be replaced in the rule by the explicit hypothesis that "all possible probabilities [are] equally likely," and noted that:

Though the cases are very rare in which the radical assumption of the Rule of Succession is strictly justified, the rule may be taken to afford a rough and ready estimate in many cases in which the assumption is approximately justified. [Whitworth, 1901, page 193]

This defense essentially originates with Edgeworth, who was an important defender of inverse methods throughout this period (see Stigler, 1978, page 296; 1986, page 310). In 1884, at the beginning of his career, Edgeworth wrote a review of Venn's *Logic*, entitled "The Philosophy of Chance," which appeared in the English philosophical journal *Mind*. (Nearly 40 years later, in the twilight of his career, Edgeworth would return to the same subject with an article of the same title in the same journal, this time reviewing Keynes's *Treatise*.) Edgeworth took an empirical and pragmatic view of the subject, and, as noted earlier, may well have been responsible for many of the changes Venn made in the third edition of *The Logic of Chance*.

The defenses mounted by Edgeworth and others fell into three broad categories. They were: (1) *The Bayes-Laplace postulate of equiprobability corresponds, at least approximately, to experience.* Karl Pearson found this argument particularly persuasive, and adopted it in his influential *Grammar of Science* (1st edition, 1892) and later articles (Pearson, 1907; Pearson, 1920, page 4). (2) *Other priors.* Another move was to concede that experience might indeed point to other priors. Both the actuary G. F. Hardy (1889) and the mathematician Whitworth (1897, pages 224–225) proposed the class of beta priors as suitable for this purpose. Others, such as Gosset (1908) and Bachelier (1912), suggested the use of polynomial priors. (3) *The suppression of a priori probabilities* (Edgeworth, 1922, page 264). A third and final defense was that when large samples were involved the particular prior employed did not matter. This had been noted as early as 1843 by both Cournot (1843, Section 95, page 170) and Mill (1843, Book 3, Chapter 18, Section 6), and had been extended by Edgeworth to parameters other than binomial proportions (Edgeworth, 1884b, page 204). A related development was Poincaré's method of arbitrary functions; see, e.g., Borel (1965, Chapter 9).

These were creditable arguments and, given the *imprimatur* of Edgeworth and Pearson, it is not surprising to find acceptance of prior probabilities at least initially even among statisticians of Fisher's own generation. Gosset's ["Student"] discussion of the issue in his classic 1908 paper on the "Probable error of a correlation coefficient" is a good example. Gosset describes the estimation problem for the correlation coefficient as that of determining "the probability that R [the population correlation coefficient] for the population from which the sample is drawn shall lie between any given limits" (Gosset, 1908, page 302). He then adds:

It is clear that in order to solve this problem we must know two things: (1) the distribution of values of r [the sample correlation coefficient] derived from samples of a population which has a given R, and (2) the *a priori* probability that R for the population lies between any given limits. Now (2) can hardly ever be known, so that some arbitrary assumption must in general be made . . . I may suggest two more or less obvious distributions. The first is that any value is equally likely between $+1$ and -1, and the second that the probability that x is the value is proportional to $1 - x^2$: this I think is more in accordance with ordinary experience: the distribution of *a priori* probability would then be expressed by the equation $y = (^3\!/_4)(1 - x^2)$.

Gosset's discussion clearly reflects a change in climate; "some arbitrary assumption must in general be made;" and a nonuniform prior seems "more in accordance with ordinary experience." Nevertheless, his basic view of estimation is clearly Bayesian. Nor were the references to prior probabilities in the statistical literature of this period mere lip-service: Edgeworth's important 1908 papers on maximum likelihood were based in part on them, and Neyman himself later employed prior probabilities in some of his earlier papers (Neyman and Pearson, 1928; Neyman, 1929). (Neyman had originally hoped to have Pearson's name appear as a co-author on the second paper, but by this time Pearson was unwilling to have his name associated in print with prior probabilities (Reid, 1982, pages 82–85).)

Acceptance of inverse methods continued into the 1920s, when they received a powerful assist from the work of Frank Ramsey (1926). Indeed, Fisher would appear to be the *first* British statistician of any standing to publicly attack Bayesian methods. The remarkably hostile reaction to his 1935 *JRSS* discussion paper (Fisher, 1935) may reflect in large part the antagonism of the Bayesian old-guard to the *nouvelle statistique*. Writing as late as 1934, Neyman could state that "until recently" it had been assumed that the problem of statistical estimation in sampling from a population required "knowledge of probabilities a priori" (Neyman, 1934).

Nearly half a century elapsed between the appearance of the first edition of Chrystal's *Algebra* (1886) and Fisher's attacks on inverse probability. During that period inverse methods were debated, claims for the theory qualified, and caution in its use advised, but the theory itself was never totally abandoned, and there is no evidence whatever for what Fisher described on one occasion as an abrupt and dramatic change. Textbooks continued to cover the subject (e.g., Coolidge, 1925; Burnside, 1928; Fry, 1928), questions on it continued to appear on actuarial examinations (A. Fisher, 1915, page 56), respected statisticians continued to employ it (Bowley, 1926). Fisher suggests that the most important result of Venn's criticism had been Chrystal's omission of inverse probability from his *Algebra*. Surely more to the point is that virtually every textbook in probability written in English during the period 1886–1930 *includes* the topic, as well as most texts in French and German. Indeed, it is difficult to find exceptions – apart from Bertrand – at least among texts of the first rank. Writing in 1921, Keynes could state that "the reaction against the traditional teaching during the past hundred years has not possessed sufficient force to displace the established doctrine, and the Principle of Indifference is still very widely accepted in an unqualified form" (Keynes, 1921, page 84).

Fisher was, in fact, being too modest when he ascribed the demise of inverse probability to Boole, Venn and Chrystal. The two most important persons in that undertaking were none other than Fisher himself and Neyman. (Thus for Egon Pearson, the inverse probability approach "had been forever discredited by Fisher in his 1922 paper . . . " (Reid, 1982, page 79).) Human nature being what it is, no matter how cogent or convincing the arguments of the opponents of inverse probability were, until a credible alternative to the Bayesian methodology was provided, any attempt to demolish the edifice of inverse probability was doomed to failure (e.g., Pearson, 1920, page 3).

The Harvard mathematician Julian Lowell Coolidge was perhaps merely being more candid than most when he wrote (1925, page 100):

Why not, then, reject the formula outright? Because, defective as it is, Bayes' formula is the only thing we have to answer certain important questions which do arise in the calculus of probability. . . . Therefore we use Bayes' formula with a sigh, as the only thing available under the circumstances:

'Steyning tuk him for the reason the thief tuk the hot stove – bekaze there was nothing else that season.' [Kipling, *Captains Courageous*, Chapter 6]

6. DISCUSSION

Paradoxically, the history of science when written by scientists themselves is sometimes seriously flawed. A typology of possible reasons for this suggests two general categories, involving sins of omission and sins of commission.

First and foremost, there may be simply a lack of interest, resources, time or training. A common manifestation of this is the uncritical copying of earlier, secondary, often highly flawed accounts without consulting original sources.

Everyone "knows," for example, that during the Middle Ages the Ptolemaic model of the solar system was modified by the addition of epicycle upon epicycle to artificially force agreement with increasingly accurate experimental data. But in reality, nothing of the kind occurred: the original Ptolemaic model of one deferent and one epicycle provided a remarkably good fit to the observational data available prior to the time of Tycho Brahe; indeed, given the mathematical sophistication of Ptolemy's original system, more simplified models were typically employed throughout the Middle Ages, not more complex ones (e.g., Gingerich, 1973, page 95). But this misconception fits popular prejudices about the science of the Middle Ages (see, e.g., Arthur Koestler's *The Sleepwalkers*, 1959) and so is repeated from one misinformed source to another. It does not occur to someone to check the authenticity of such a story, any more than it would occur to him to check whether Einstein was responsible for the special theory of relativity, or whether Watson and Crick discovered the structure of DNA.

Even when a person has first-hand knowledge of the events about which he is writing, the passage of time may lead to a subtle erosion in the accuracy with which those events are remembered. A notable example is Karl Pearson's historical account of correlation (Seal, 1967; Plackett, 1983). As Stigler notes, Pearson's commentary "reflects well neither upon Pearson nor the general trustworthiness of the latter recollections of great scientists" (Stigler, 1986, page 344, n. 11).

Under the rubric of sins of commission may be placed an interrelated complex of causes including subconscious bias, dogmatism, sensationalism and deliberate distortion. Everyone "knows," for example, that the night before he was fatally wounded in a duel, the unfortunate Évariste Galois stayed up feverishly writing down a sketch of his theory of equations so that it would not be lost to posterity. In reality Galois had published an outline of his results months earlier, and although he did write further details down the night before the fatal duel, there was not the urgency often depicted. Reality does not make nearly as good a story as the piquant version in circulation. As Rothman (1982) discusses, this is not an isolated incident in Galois's biography: several of the best known accounts of Galois's life (those of Bell, Hoyle and Infeld) are marred by serious inaccuracies which occur because of – rather than in spite of – the ability of their authors to appreciate the technical achievements of Galois; "the misfortune is that the biographers have been scientists" (Rothman, 1982, page 104). Similarly, Stigler (1982) argues that many accounts of Bayes's original paper are seriously inaccurate; here foundational biases often led statisticians of the stature of Pearson, Fisher and Jeffreys to misread into Bayes their own viewpoints.

Fisher's account of the history of inverse probability is marred for reasons falling into both of these general categories. Due perhaps in part to poor eyesight, Fisher was never very scholarly in documenting previous work; this was to prove vexatious years later when Neyman and others would criticize him for not adequately acknowledging Edgeworth's earlier contributions to maximum likelihood (Savage, 1976, pages 447–448; Pratt, 1976).

Nevertheless, throughout his life Fisher had a serious interest in historical matters. Leafing through Todhunter, he was quick to note the Bernoulli-Montmort correspondence about the optimal strategy in the game of "le Her," and realized (a decade before the work of von Neumann and Morgenstern on game theory) that a randomized strategy was appropriate (Fisher, 1934). (On the other hand, had Fisher referred to Montmort's book he would have discovered an extract of a letter from Waldegrave to Montmort discussing the possibility of randomized strategies! (Montmort, 1713, pages 409–412).) He was often fond of using an historical data set as the perfect pedagogical foil; the entire third chapter of Fisher's *Design of Experiments*, for example, is centered about an analysis of Darwin's data on cross and self-fertilized plants. Occasionally, the result might even suggest a radical historical reassessment, as in his article on whether Mendel fudged his data (Fisher, 1936a; Root-Bernstein, 1983).

And what Fisher was acquainted with, he often knew very well indeed. As Savage (1976, page 447) notes, Fisher "was well read in the statistical literature of his past," and Fisher's writings display a detailed knowledge of Bayes, Boole, Venn, Todhunter and Keynes. But it is a common failing to read into the words of the past the thoughts of the present, and to view the evolution of history as the progressive triumph of one's own viewpoint. This Fisher appears to have done.

ACKNOWLEDGMENTS

The author expresses his thanks to Persi Diaconis and Paul Meier for a number of helpful comments and suggestions during the preparation of the paper, to Elisabeth Vodola for supplying a copy of Chrystal's 1891 paper, and to an anonymous referee for a careful reading of the manuscript.

REFERENCES

Bachelier, L. (1912). *Calcul des probabilités* **1**. Paris: Gauthier-Villars.
Bertrand, J. (1889). *Calcul des probabilités*. Paris: Gauthier-Villars. (2nd ed., 1907, reprinted by New York: Chelsea.)

Boole, G. (1854). *An Investigation of the Laws of Thought.* London: Walton and Maberly. (Reprinted by Dover, New York, 1976.)

Boole, G. (1862). On the theory of probabilities. *Philos. Trans. Roy. Soc. London* **152** 225–252. (Reprinted in G. Boole, *Collected Logical Works* **1**. *Studies in Logic and Probability* (R. Rhees, ed.) 386–424. La Salle, Ill.: Open Court Publishing Co., 1952.)

Borel, É. (1965). *Elements of the Theory of Probability.* Englewood Cliffs, N.J.: Prentice-Hall.

Bowley, A. L. (1926). Measurement of the precision of index-numbers attained in sampling. *Bull. Internat. Statist. Inst.* **22** 6–62.

Box, J. F. (1978). *R. A. Fisher: The Life of a Scientist.* New York: Wiley.

Burnside, W. (1928). *The Theory of Probability.* New York: Cambridge University Press. (Reprinted by Dover, New York, 1959.)

Carnap, R. (1950). *Logical Foundations of Probability.* Chicago: University of Chicago Press. (2nd ed., 1962).

Chrystal, G. (1886). *Algebra.* London: Adam and Charles Black.

Chrystal, G. (1891). On some fundamental principles in the theory of probability. *Trans. Actuarial Soc. Edinburgh (N. S.)* **2** 421–439.

Coolidge, J. L. (1925). *An Introduction to Mathematical Probability.* Oxford University Press. (Reprinted by Dover, New York, 1962.)

Cournot, A. A. (1843). Exposition de la théorie des chances et des probabilités. Librairie de L. Hachette, Paris.

De Finetti, B. (1972). *Probability, Induction, and Statistics: The Art of Guessing.* New York: Wiley.

De Morgan, A. (1847). *Formal Logic: or, the Calculus of Inference, Necessary and Probable.* London: Taylor and Walton. (Reprinted by The Open Court Co., London, 1926.)

Donkin, W. F. (1851). On certain questions relating to the theory of probabilities. *Philos. Mag. (4)* **1** 353–368, 458–466.

Edgeworth, F. Y. (1884a). The philosophy of chance. *Mind* **9** 222–235.

Edgeworth, F. Y. (1884b). *A priori* probabilities. *Philos. Mag. (5)* **18** 204–210.

Edgeworth, F. Y. (1908). On the probable errors of frequency constants. *J. Roy. Statist. Soc.* **71** 381–397, 499–512, 651–678. Addendum **72** (1909), 81–90.

Edgeworth, F. Y. (1922). The philosophy of chance. *Mind* **31** 257–283.

Ellis, R. L. (1844). On the foundations of the theory of probabilities. *Trans. Cambridge Philos. Soc.* **8** 1–6. (Reprinted in *The Mathematical and Other Writings of Robert Leslie Ellis M. A.* (W. Walton, ed.). Cambridge, UK: Deighton and Bell, 1863.)

Feinberg, S. E. and Hinkley, D. V. (eds.) (1980). *R. A. Fisher: An Appreciation. Lecture Notes in Statist* **1**. New York: Springer.

Fisher, A. (1915). *The Mathematical Theory of Probabilities and its Application to Frequency Curves and Statistical Methods* **1**. *Mathematical Probabilities and Homograde Statistics*, 2nd ed. New York: Macmillan, 1923.

Fisher, R. A. (1912). On an absolute criterion for fitting frequency curves. *Messenger Math.* **41** 155–160. (*Collected Papers* **1**.)

Fisher, R. A. (1921). On the "probable error" or a coefficient of correlation deduced from a small sample. *Metron* **1** 3–32. (*Collected Papers* 14; contains Fisher's first critical comment on inverse probability.)

Fisher, R. A. (1922). On the mathematical foundations of theoretical statistics. *Philos. Trans. Roy. Soc. London Ser. A* **222** 309–368. (*Collected Papers* 18.)

Fisher, R. A. (1930). Inverse probability. *Proc. Cambridge Philos. Soc.* **26** 528–535. (*Collected Papers* 84.)

Fisher, R. A. (1934). Randomisation, and an old enigma of card play. *Math. Gaz.* **18** 294–297. (*Collected Papers* 111.)

Fisher, R. A. (1935). The logic of inductive inference. *J. Roy. Statist. Soc.* **98** 39–54. (*Collected Papers* 124.)

Fisher, R. A. (1936a). Uncertain inference. *Proc. Amer. Acad. Arts Sci.* **71** 245–258. (*Collected Papers* 137.)

Fisher, R. A. (1936b). Has Mendel's work been rediscovered? *Ann. Science* **1** 115–137. (*Collected Papers* 144.)

Fisher, R. A. (1951). Statistics. In *Scientific Thought in the Twentieth Century* (A. E. Heath, ed.) 31–55. London: Watts. (*Collected Papers* 242.)

Fisher, R. A. (1955). Statistical methods and scientific induction. *J. Roy. Statist. Soc. Ser. B* **17** 69–78. (*Collected Papers* 261.)

Fisher, R. A. (1956). *Statistical Methods and Scientific Inference.* New York: Hafner. (2nd ed., 1959; 3rd ed., 1973; page references are to the 3rd ed.)

Fisher, R. A. (1958). The nature of probability. *Centennial Review* **2** 261–274. (*Collected Papers* 272.)

Fisher, R. A. (1971–74). *Collected Papers of R. A. Fisher* **1–5** (J. H. Bennett, ed.). University of Adelaide.

Fry, T. C. (1928). *Probability and Its Engineering Applications.* New York: van Nostrand.

Gingerich, O. (1973). Copernicus and Tycho. *Scientific American* **229** 86–101.

Gosset, W. S. (1908). Probable error of a correlation coefficient. *Biometrika* **6** 302–310.

Hailperin, T. (1976). *Boole's Logic and Probability.* Amsterdam: North-Holland.

Hardy, G. F. (1889). Letter. *Insurance Record* 457. (Reprinted, *Trans. Faculty Actuaries* **8** 180–181, 1920.)

Jeffreys, H. (1939). *Theory of Probability.* Oxford, UK: Clarendon Press. (2nd ed., 1948; 3rd ed., 1967.)

Jevons, W. S. (1877). *The Principles of Science*, 2nd ed. London: Macmillan.

Keynes, J. M. (1921). *A Treatise on Probability.* London: Macmillan.

Koestler, A. (1959). *The Sleepwalkers.* New York: Macmillan.

Mill, J. S. (1843). *A System of Logic, Ratiocinative and Inductive, Being a Connected View of the Principles of Evidence and the Methods of Scientific Investigation.* London: John W. Parker. (Many later editions.)

Montmort, P. R. (1713). *Essai d'analyse sur les jeux de hazards*, 2nd ed. Paris: Jacques Quillan. (1st ed., 1708.)

Neyman, J. (1929). Contribution to the theory of certain test criteria. *Bull. Internat. Statist. Inst.* **24** 3–48.

Neyman, J. (1934). On the two different aspects of the representative method: The method of stratified sampling and the method of purposive selection. *J. Roy. Statist. Soc.* **97** 558–625.

Neyman, J. and Pearson, E. S. (1928). On the use of interpretation of certain test criteria for purposes of statistical inference. *Biometrika* **20** 175–240, 263–294.

Passmore, J. (1968). *A Hundred Years of Philosophy*, 2nd ed. New York: Penguin.

Pearson, K. (1892). *The Grammar of Science*. Walter Scott, London. (2nd ed., 1900; 3rd ed., 1911.)

Pearson, K. (1907). On the influence of past experience on future expectation. *Philos. Mag.* (6) **13** 365–378.

Pearson, K. (1920). The fundamental problem of practical statistics. *Biometrika* **13** 1–16.

Plackett, R. L. (1983). Karl Pearson and the chi-squared test. *Internat. Statist. Rev.* **51** 59–72.

Porter, T. M. (1986). *The Rise of Statistical Thinking: 1820–1900*. Princeton, N.J.: Princeton University Press.

Pratt, J. W. (1976). F. Y. Edgeworth and R. A. Fisher on the efficiency of maximum likelihood estimation. *Ann. Statist.* **4** 501–514.

Ramsey, F. P. (1926). Truth and Probability. In *The Foundations of Mathematics and Other Logical Essays* (R. B. Braithwaite, ed.) 156–198. London: Routledge and Kegan Paul, (1931).

Reid, C. (1982). *Neyman – From Life*. New York: Springer.

Root-Bernstein, R. S. (1983). Mendel and methodology. *History of Science* **21** 275–295.

Rothman, T. (1982). Genius and biographers: the fictionalization of Évariste Galois. *Amer. Math. Monthly* **89** 84–106.

Salmon, W. C. (1981). John Venn's *Logic of Chance*. In *Probabilistic Thinking, Thermodynamics and the Interaction of the History and Philosophy* (J. Hintikka, D. Gruender and E. Agazzi, eds.) **2** 125–138. Dordrecht: Reidel.

Savage, L. J. (1976). On re-reading R. A. Fisher (with discussion). *Ann. Statist.* **3** 441–500.

Seal, H. L. (1967). The historical development of the Gauss linear model. *Biometrika* **54** 1–24.

Shafer, G. (1976). *A Mathematical Theory of Evidence*. Princeton, N.J.: Princeton University Press.

Stigler, S. M. (1978). Francis Ysidro Edgeworth, statistician (with discussion). *J. Roy. Statist. Soc. Ser. A* **141** 287–322.

Stigler, S. M. (1982). Thomas Bayes's Bayesian inference. *J. Roy. Statist. Soc. Ser. A* **145** 250–258.

Stigler, S. M. (1986). *The History of Statistics: The Measurement of Uncertainty Before 1900*. Cambridge, Mass.: Harvard University Press.

Todhunter, I. (1865). *A History of the Mathematical Theory of Probability*. London: Macmillan. (Reprinted by Chelsea, New York, 1949.)

Venn, J. (1866). *The Logic of Chance*. London: Macmillan. (2nd ed., 1876; 3rd ed., 1888; reprinted by Chelsea, New York, 1962.)

von Kries, J. (1886). *Die Prinzipien der Wahrscheinlichkeits-rechnung. Eine Logische Untersuchung*. Freiburg. (2nd ed., Tübingen, 1927.)

von Wright, G. H. (1941). *The Logical Problem of Induction*. Finnish Literary Soc., Helsinki. (2nd rev. ed. New York: Macmillan, 1957.)

Whittaker, E. T. (1920). On some disputed questions of probability (with discussion). *Trans. Faculty Actuaries* **77** 163–206.

Whitworth, W. A. (1897). *DCC Exercises in Choice and Chance*. (Reprinted by Hafner, New York, 1965.)

Whitworth, W. A. (1901). *Choice and Chance*, 5th ed. London: George Bell and Sons.

Winsor, C. P. (1947). Probability and listerism. *Human Biology* **19** 161–169.

8

R. A. Fisher and the Fiducial Argument

Abstract. The fiducial argument arose from Fisher's desire to create an inferential alternative to inverse methods. Fisher discovered such an alternative in 1930, when he realized that pivotal quantities permit the derivation of probability statements concerning an unknown parameter independent of any assumption concerning its a priori distribution.

The original fiducial argument was virtually indistinguishable from the confidence approach of Neyman, although Fisher thought its application should be restricted in ways reflecting his view of inductive reasoning, thereby blending an inferential and a behaviorist viewpoint. After Fisher attempted to extend the fiducial argument to the multiparameter setting, this conflict surfaced, and he then abandoned the unconditional sampling approach of his earlier papers for the conditional approach of his later work.

Initially unable to justify his intuition about the passage from a probability assertion about a statistic (conditional on a parameter) to a probability assertion about a parameter (conditional on a statistic), Fisher thought in 1956 that he had finally discovered the way out of this enigma with his concept of *recognizable subset*. But the crucial argument for the relevance of this concept was founded on yet another intuition – one which, now clearly stated, was later demonstrated to be false by Buehler and Feddersen in 1963.

Key words and phrases: Fiducial inference, R. A. Fisher, Jerzy Neyman, Maurice Bartlett, Behrens-Fisher problem, recognizable subsets.

Most statistical concepts and theories can be described separately from their historical origins. This is not feasible, without unnecessary mystification, for the case of "fiducial probability." (Stone, 1983, p. 81)

1. INTRODUCTION

Fiducial inference stands as R. A. Fisher's one great failure. Unlike Fisher's many other original and important contributions to statistical methodology

Reprinted with permission from *Statistical Science* 7, no. 3 (1992): 369–387.

and theory, it has never gained widespread acceptance, despite the importance that Fisher himself attached to the idea. Instead, it was the subject of a long, bitter and acrimonious debate within the statistical community, and while Fisher's impassioned advocacy gave it viability during his own lifetime, it quickly exited the theoretical mainstream after his death.

Considerable confusion has always existed about the exact nature of the fiducial argument, and the entire subject has come to have an air of mystery about it. The root causes of such confusion stem from several factors. First and foremost, Fisher's own thoughts on fiducial inference underwent a substantial evolution over time, and both a failure on his part to clearly acknowledge this and a failure by others to recognize such changes have often led to confusion (when attempting to reconcile conflicting passages in Fisher's writings), or misinterpretation (when a later position is misread into an earlier). Second, fiducial inference never actually developed during Fisher's lifetime into a coherent and comprehensive theory, but always remained essentially a collection of examples, insights and goals, which Fisher added to and modified over time. Viewed in this limited way, the "theory" becomes at once much less ambitious and much more credible. Finally, the polemical nature of the debate on both sides rendered much of the resulting literature opaque: neither side was willing to concede inadequacies or limitations in its position, and this often makes any single paper difficult to understand when read in isolation.

This paper attempts to trace the roots and evolution of Fisher's fiducial argument by a careful examination of his own writings on the subject over a period of some thirty years. As will be seen, Fisher's initial insight and basic goals throughout are readily understood. But his attempts to extend the argument to the multiparameter setting and the criticism of his views by others led Fisher to reformulate the initial fiducial argument, and the approach taken in his later papers is very different from that to be found in his writings two decades earlier.

Although the last section of this paper briefly comments on the various efforts made after Fisher's death to clarify, systematize and defend the fiducial argument, our primary interest is what Fisher himself did (or did not) accomplish.

There are several "theses" advanced, stated below. These serve the useful purpose of summarizing the ensuing argument but necessarily omit a number of qualifications discussed later. Specifically, we will argue the following:

- Fisher's opposition to Bayesian methods arose (at least in part) from his break with Pearson; fiducial inference was intended as an "objective" alternative to "subjective," arbitrary Bayesian methods.

- Fisher's original fiducial argument was radically different from its later versions and was largely indistinguishable from the unconditional confidence interval approach later championed by Neyman.
- In response to Neyman's confidence interval formulation, Fisher drew attention to the multiplicity of conflicting parameter estimates arising from that approach, and in an attempt to deal with this difficulty he then explicitly imposed a further condition necessary for the application of the fiducial argument.
- As a result of his debate with Bartlett, Fisher became increasingly concerned with the conditional nature of inference, and this led to a dramatic shift in his conception of fiducial inference.
- Sensing that he was fighting a losing battle in the middle 1950s, Fisher made a supreme effort to spell out as clearly as he could the nature of the fiducial argument. In doing so, however, he revealed that the new intuitions he had about the fiducial argument were fundamentally incoherent.

2. FROM INVERSE TO FIDUCIAL PROBABILITY

Fisher began life a Bayesian. He tells us that, while at school, he learned the theory of inverse probability "as an integral part of the subject, and for some years saw no reason to question its validity" (Fisher, 1936, p. 248); he pled guilty to having, in his very first paper, "based my argument upon the principle of inverse probability" (Fisher, 1922, p. 326), and he thought it worth noting from an historical standpoint that "the ideas and nomenclature for which I am responsible were developed only after I had inured myself to the absolute rejection of the postulate of Inverse Probability" (CP 159A, p. 151).

Fisher saw fiducial inference as the jewel in the crown of the "ideas and nomenclature" for which he was responsible,[1] and in order to appreciate what he intended to achieve with it, we may perhaps best begin by considering what led him to so decisively reject the methods of inverse probability in the first place.

In 1915, Fisher published his first major paper in statistics, in which he derived the exact distribution of the sample correlation coefficient (Fisher, 1915). Although this paper was published in Karl Pearson's journal *Biometrika*, two years later a "cooperative study" by Pearson and several associates appeared criticizing Fisher's paper on several grounds (Soper et al., 1917). One of these, which particularly annoyed Fisher, was the (erroneous) charge that he had employed a Bayesian solution with an inappropriate prior for the correlation coefficient ρ.[2]

Relations between Fisher and Pearson rapidly worsened: by 1918 Pearson had rejected as referee Fisher's later famous paper on the correlation of relatives (Fisher, 1918), and the next year Fisher refused an offer from Pearson to join his laboratory, going to Rothamsted instead (Box, 1978, pp. 61, 82–83). Despite this, in 1920 Fisher again submitted a paper to Pearson for publication in *Biometrika*, but when this too was rejected Fisher vowed he would never do so again (Box, 1978, p. 83).[3]

Fisher's animosity toward Pearson is well known, but to gauge the true depth of his anger it is instructive to read the bitter Foreword to his book *Statistical Methods and Scientific Inference* (Fisher, 1956), written almost twenty years after Pearson's death.[4] It is at least arguable that in some cases the direction Fisher's statistical research now took – and the manner in which his papers were written – were motivated in part by a desire to attack Pearson. After moving to Rothamsted, Fisher proceeded (in a series of five papers published over the seven-year period 1922–1928) to attack Pearson's use of the chi-squared statistic to test homogeneity, on the (entirely correct) grounds that Pearson had systematically employed an incorrect number of degrees of freedom (Box, 1978; Feinberg, 1980). At the same time, Fisher began to publish his landmark papers on estimation. Although criticism of Pearson's work was not central to these, a key element of Fisher's new theory was the notion of efficient methods of estimation, and Fisher was quick to point out that Pearson's method of moments was frequently inefficient.

Pearson was also an exponent of Bayesian methods, and thus Fisher's rejection of inverse methods and his development of fiducial inference as an alternative to them was yet another assault on the Pearsonian edifice.[5] Less than a year after the rejection of his 1920 paper, Fisher fired the first salvo, asserting that the approach taken by Bayes "depended upon an arbitrary assumption, so that the whole method has been widely discredited" (Fisher, 1921, p. 4) and pointed an accusing finger at "inverse probability, which like an impenetrable jungle arrests progress towards precision of statistical concepts" (Fisher, 1922, p. 311).

Fisher could write with considerable conviction about the arbitrary nature of Bayesian analyses, for he felt that he had been one of its most recent victims. The writers of the *Cooperative Study*, Fisher charged, had altered

my method by adopting what they consider to be a better *a priori* assumption as to ρ. This they enforce with such rigor that a sample which expresses the value 0.600 has its message so modified in transmission that it is finally reported as 0.462 at a distance of 0.002 only above that value which is assumed *a priori* to be most probable! (Fisher, 1921, p. 17)

The resulting value, Fisher thus noted, "depends almost wholly upon the preconceived opinions of the computer and scarcely at all upon the actual data supplied to him."

The close relationship between the *Cooperative Study* episode, and Fisher's subsequent and vehement rejection of inverse methods is evident in Fisher's original paper on fiducial inference (Fisher, 1930), appropriately called "Inverse Probability." Although the simplest examples of fiducial intervals would have been those for the mean and standard deviation, they were not employed by Fisher, who used instead the more complex example of the fiducial interval for the correlation coefficient, that is, precisely the setting in which Pearson had dared to criticize Fisher 13 years earlier for employing an inappropriate Bayesian solution. The slight had not been forgotten.

The exchange with Pearson impressed on Fisher the arbitrariness and dangers inherent in the use of priors lacking empirical support. By 1930, however, Fisher believed that he had discovered a way out of this difficulty, by employing what he termed the *fiducial argument*.

3. THE (INITIAL) FIDUCIAL ARGUMENT

3.1. The Birth of the Fiducial Argument

The fiducial argument was born during conversations between Fisher and his colleagues at Rothamsted.[6] A key role was played by the biologist E. J. Maskell, who worked there in the early 1920s. Maskell made the simple but important observation that when estimating the mean of a population, one could, in place of the usual two standard error limits, equally well employ the percentiles of the t-distribution to derive interval estimates corresponding to any desired level of significance.

Fisher briefly alluded to Maskell's role in Chapter 10 of the *Design of Experiments* (Fisher, 1935c). Referring to the classical example of Darwin's paired comparison of the heights of cross- and self-fertilized plants (introduced earlier in Chapter 3 of *Design of Experiments*, for which $n = 15$, $\bar{x} = 20.933$, $s = 37.744$, and $s/\sqrt{15} = 9.746$, Fisher wrote:

An important application, due to Maskell, is to choose the values of t appropriate to any chosen level of significance, and insert them in the equation. Thus t has a 5 per cent. chance of lying outside the limits ± 2.154. Multiplying this value by the estimated standard deviation, 9.746, we have 20.90 and may write

$$\mu = 20.93 \pm 20.90 = 0.03, \text{ or } 41.83$$

as the corresponding limits for the value of μ.[7]

But although Fisher thus knew the substance of the fiducial argument no later than 1926 (when Maskell left Rothamsted for the Caribbean), he did not refer to it in print for several years, perhaps because the initial observation, tied to the special case of the *t*-distribution, seemed too simple to warrant publication. But this changed by 1930, when Fisher discovered a way of generalizing the argument to cover a large class of univariate parameter estimates.

3.2. *"Inverse Probability"*

It is in many ways ironic that Fisher's first paper on fiducial inference, entitled "Inverse Probability" (Fisher, 1930), contains little that would be considered controversial today. In it Fisher introduced the probability integral transformation and observed that this transformation often provides a pivotal quantity which may be inverted to obtain interval estimates having any prespecified coverage frequency. That is, Fisher not only gave a clear and succinct statement of (what later came to be called) the confidence interval approach to parameter estimation, but (and this appears almost universally unappreciated) he also gave a general method for obtaining such estimates in the one-dimensional case.[8]

Fisher specifically observed that if a continuous statistic T exists whose sampling distribution "is expressible solely in terms of a single parameter" θ, then T can often be inverted to obtain probability statements about θ which are true "irrespective of any assumption as to its *a priori* distribution":

If T is a statistic of continuous variation, and P the probability that T should be less than any specified value, we have then a relation of the form

$$P = F(T, \theta).$$

If now we give to P any particular value such as 0.95, we have a relationship between the statistic T and the parameter θ, such that T is the 95 per cent. value corresponding to a given θ, and this relationship implies the perfectly objective fact that in 5 per cent. of samples T will exceed the 95 per cent. value corresponding to the actual value of θ in the population from which it is drawn. To any value of T there will moreover be usually a particular value of θ to which it bears this relationship; we may call this the "fiducial 5 per cent. value of θ" corresponding to a given T. If, as usually if not always happens, T increases with θ for all possible values, we may express the relationship by saying that the true value of θ will be less than the fiducial 5 per cent. value corresponding to the observed value of T in exactly 5 trials in 100. By constructing a table of corresponding values, we may know as soon as T is calculated what is the fiducial 5 per cent. value of θ, and that the true value of θ will be less than this value in just 5 per cent. of trials. This then is a definite probability statement about the unknown parameter θ which is true irrespective of any assumption as to its *a priori* distribution. (Fisher, 1930, pp. 532–533)

That is, if $F(t, \theta) =: P_\theta[T \leq t]$, and if for each $p \in [0,1]$, the relation $F(t, \theta) = p$ implicitly defines functions $\theta_p(t)$ and $t_p(\theta)$ such that (i) $F(t_p(\theta), \theta) = p$ and (ii) $\theta_p(t) \leq \theta \Leftrightarrow t \leq t_p(\theta)$, then

$$P_\theta[\theta_p(T) \leq \theta] = p$$

whatever the value of θ. Fisher termed $\theta_p(t)$ the "fiducial" $100(1 - p)$ percent value corresponding to t.[9]

This simple mathematical observation cannot, of course, be faulted, and all subsequent controversy about the fiducial argument has centered around either the *interpretation* of this result or the attempt to extend the argument to other contexts (discontinuous or multiparameter). Let us consider some of the issues raised either by Fisher or others.

3.3. The Interpretation of a Fiducial Probability

At this initial stage, Fisher's interpretation of the "fiducial" probability state-ment, as the quotation above makes clear, was closely tied to frequency con-siderations and coverage properties. Nor was the language occurring here an isolated instance, for it is closely paralleled by the language Fisher used in his next paper concerning the fiducial argument (Fisher, 1933).[10]

It might be argued that Fisher intended such references to frequency as simply stating a property (one among many) enjoyed by a fiducial interval, rather than being an essential element in its definition. Such an interpretation, however, is not supported by Fisher's language, for he went on to add (referring to the example of the correlation coefficient):

[I]f a value $r = 0.99$ were obtained from the sample, we should have a fiducial 5 per cent. ρ equal to about 0.765. The value of ρ can then only be less than 0.765 in the event that r has exceeded its 95 per cent. point, an event which is known to occur just once in 20 trials. In this sense ρ has a probability of just 1 in 20 of being less than 0.765. (Fisher, 1930, p. 534)

"In this sense" – this crucial phrase makes it clear that the references to sampling frequency that occur here and elsewhere were *central* to Fisher's conception of fiducial probability at this stage, not subsidiary to it.[11] The use of the adjective "fiducial," as Fisher himself repeatedly emphasized, was intended only to underscore the novel mode of derivation employed and was

not meant to suggest that a new and fundamentally different type of probability was involved (as *was* the case with the distinction Fisher drew between
probability and likelihood).[12]

3.4. The Fiducial Distribution

If the function $G(p) =: \theta_p(t)$ is strictly decreasing in p, then its inverse $G^{-1}(\theta)$
is certainly a distribution function in the mathematical sense that $H(\theta) =: 1 - G^{-1}(\theta)$ is a continuous increasing function with $H(-\infty) = 0$ and $H(+\infty) = 1$;
Fisher termed it "the fiducial distribution" of the parameter θ corresponding
to the value t and noted that it has the density $-\partial F(t, \theta)/\partial\theta$.

Fisher regarded this result as supplying "definite information as to the
probability of causes" and viewed the fiducial distribution as a probability
distribution for θ in the ordinary sense. This is made clear at the end of the
1930 paper, when Fisher contrasted the fiducial and inverse approaches. At
this stage, Fisher thought the fiducial argument was valid *even when* a prior
distribution for θ was known. Because the resulting posterior and fiducial
distributions ordinarily differ, Fisher stressed that they were really saying
very different things: that although both were probability distributions, their
"logical meaning" or "content" differed (Fisher, 1930, p. 534; 1933, pp. 82–
83; 1936, p. 253), a position he later disavowed, for reasons that will be
discussed below.[13]

Indeed Fisher's (1930) discussion (once again appealing to the correlation
coefficient example he had used earlier) reveals just how unconditional a
sampling interpretation he held at this juncture:

In concrete terms of frequency this would mean that if we repeatedly selected a
population at random, and from each population selected a sample of four pairs of
observations, and rejected all cases in which the correlation as estimated from the
sample (r) was not exactly 0.99, then of the remaining cases 10 per cent. would have
values of ρ less than 0.765. Whereas apart from any sampling for ρ, we know that
if we take a number of samples of 4, from the same or different populations, and for
each calculate the fiducial 5 per cent. value for ρ, then in 5 per cent. of cases the true
value of ρ will be less than the value we have found. There is thus no contradiction
between the two statements. (p. 535)

Little wonder that many statisticians during the 1930s regarded Fisher's
theory of fiducial inference and Neyman's theory of confidence intervals as
virtually synonymous![14] But despite the close similarities between the fiducial
argument that Fisher presented in 1930 and Neyman's subsequent theory,

there was – even at this early stage – an important difference in emphasis between the two. Every confidence interval is equivalent to a series of tests of significance, and it is clearly this second interpretation that Fisher had in mind. [Dempster (1964) expresses a similar philosophy, noting that "a particular 95% confidence region determined by observed data is simply the set of parameter values *not surprising* at the 0.05 level" (p. 58), and he suggests the term *indifference region* as more appropriate.][15] Fisher remained true to this interpretation, although his later analysis of just what constitutes a valid test of significance eventually led him to largely abandon the unconditional viewpoint adopted in these earlier papers.

To summarize thus far: for every fixed value of θ, the statement $P_\theta[\theta_p(T) \leq \theta] = p$ has an unambiguous sampling interpretation for each p; for every fixed value of t, the function $H_t(\theta) =: 1 - F(t, \theta)$ is (in a purely mathematical sense) a distribution function for θ. But Fisher did not regard the resulting fiducial distribution as a probability distribution for θ *in the sense that* it describes the frequency of θ in a population having fixed values of T;[16] the fiducial distribution of θ is only one in the sense that it is the "aggregate" of the probability statements $\{P_\theta[\theta_p(T) \leq \theta] = p : 0 \leq p \leq 1\}$ (each of which refers to the frequency of T in a population having fixed values of θ).[17] Fisher wrote in 1935:

The [fiducial] *distribution* . . . is independent of all prior knowledge of the distribution of μ, *and is true of the aggregate of all samples without selection*. It involves $\bar{x}$ and s as parameters, but does not apply to any special selection of these quantities. (Fisher, 1935a, p. 392, emphasis added)

Thus, the fiducial *distribution* itself, and not just the individual probability statements $P_\theta[\theta_p(T) \leq \theta] - p$ which comprise it, must be interpreted in sampling terms. The point is that (for Fisher) every probability must be interpreted as a frequency in a population, and in this case the population is the one generated by repeated sampling: the "aggregate of all samples without selection." For $T = t$, one can compute the mathematical distribution function $H_t(\theta)$, but the probabilities in question do not refer to frequencies in a population where t is fixed and θ variable.

Nevertheless, although the point is not discussed in his 1930 paper, Fisher did regard the fiducial distribution $H_t(\theta)$, given the observed sample value $T = t$, as a numerical measure of our rational degree of belief about different possible values of θ in the light of the sample, and on at least one occasion (although only in a letter), Fisher used the fiducial distribution for a fixed value of t to compute distributional quantities such as the mean and median.[18] In

169

order to understand this apparent conundrum, we need to pause to consider Fisher's concept of probability.

3.5. The Nature of Probability

Despite the straightforward nature of Fisher's 1930 paper, his language suggests the presence of more complex and potentially inconsistent views lurking beneath the surface. On the one hand, fiducial probabilities are defined in terms of objective, unconditional sampling frequencies; on the other, the fiducial argument is said to give rise to a "probability statement *about* the unknown parameter" (Fisher, 1930, p. 533; 1933, p. 82; 1935a, p. 391). The tension arises because for Fisher a probability is – by definition – a frequency in an infinite hypothetical population (Fisher, 1922), but it is also regarded by him as a "numerical measure of rational belief" (Fisher, 1930, p. 532; see also Fisher, 1935b, p. 40; Bennett, 1990, p. 121).

Fisher nowhere gives a systematic exposition of his pre-1940 views concerning probability, but its general outlines can be deduced from the scattered comments he makes throughout his papers.[19] For Fisher, probability has an objective value:[20] it is "a physical property of the material system concerned"[21] and is independent of our state of knowledge.[22] Numerically, it is a ratio of frequencies in an infinite hypothetical population,[23] that is, a mathematical limit of frequencies in finite populations.[24]

The process of statistical inference proceeds "by constructing a hypothetical infinite population, of which the actual data are regarded as constituting a random sample" (Fisher, 1922, p. 311). Such a population, being infinite, is necessarily imaginary, a mental construct: it is the "conceptual resultant of the conditions which we are studying" (1925, p. 700) and consists of the "totality of numbers produced by the same matrix of causal conditions" (1922, p. 313). Probability is defined in terms of hypothetical frequencies, not a limit of actual experimental frequencies, because we have no knowledge of the existence of such infinite experimental limits.[25] Nevertheless, experimental frequencies are an observational measure of probability, permitting their experimental verification (Fisher, 1934, p. 4).

Thus, for Fisher a probability is a frequency, an objective property of a specified population. But probability is also epistemic: it is the basis of inductive or uncertain inferences (Fisher, 1934, p. 6), plays a role in psychological judgment (1934, p. 287) and is a "numerical measure of rational belief" (Fisher, 1930, p. 532). This passage from a frequentist denotation to an epistemic connotation is the result of an unspecified process by means of which a class-frequency can be transferred from the class to an individual in that class (Fisher, 1935b):

I mean by mathematical probability only that objective quality of the individual which corresponds to frequency in the population, of which the individual is spoken of as a typical member. (p. 78)

Thus, in the fiducial argument, given an observed value of T, say t, the probability statement concerning the parameter, $P[\theta_p(t) \leq \theta] = p$ is a numerical measure of our rational degree of belief concerning θ, originating in the statement of objective frequency regarding the statistic $\theta_p(T)$, namely $P_\theta[\theta_p(T) \leq \theta] = P_\theta[T \leq t_p(\theta)]$, but then transferred after the observation of $T = t$ to the unknown and initially nonrandom parameter θ. This can be found most clearly stated in a letter written much later to David Finney:

The frequency ratio in the entire set, therefore, is the probability of the inequality being realized in any particular case, in exactly the same sense as the frequency in the entire set of future throws with a die gives the probability applicable to any particular throw in view. (Bennett, 1990, p. 98)

But as Fisher later came to realize in 1955, this passage from the frequency for a class to an epistemic probability for an individual indeed requires some justification. (Philosophers discuss this question under the rubric of the "problem of the single-case"; for example, Reichenbach and Salmon.)

In Fisher's writings, probability often seems to live a curious Jekyll and Hyde existence; for much of the time, probability leads a quiet and respectable life as an objective frequency (Dr. Jekyll), but it occasionally transforms before our very eyes into a rational degree of belief or even a psychological mental state (Mr. Hyde, of course). For most of us, the Jekyll and the Hyde peacefully coexist, but in Stevenson's tale, a crisis arises when the two begin to struggle for supremacy. Such a drama also occurred in the case of the fiducial argument.

4. NEYMAN AND CONFIDENCE INTERVALS

4.1. Neyman's 1934 JRSS Paper

Neyman left Poland at the beginning of 1934 in order to assume a permanent academic position at University College London. Shortly after his arrival in England, Neyman read a paper before the Royal Statistical Society (on 19 June 1934) dealing in part with the fiducial argument and reformulating Fisher's theory in terms of what Neyman called "confidence intervals" (Neyman, 1934).

171

After Neyman read his paper, one of the discussants who rose to comment on it was Fisher. The exchange between the two, taking place before relations between them broke down, is instructive. The tone was polite: in introducing his theory of confidence intervals, Neyman had described it as an alternative description and development of Fisher's theory of fiducial probability, permitting its extension to the several parameter case. Fisher, ironically one of the few to comment favorably on Neyman's paper, referred to Neyman's work as a "generalization" of the fiducial argument, but pointed to the problem of a possible lack of uniqueness in the resulting probability statements if sufficient or ancillary statistics were not employed and "the consequent danger of apparently contradictory inferences."[26]

Fisher began his discussion of fiducial inference (after briefly alluding to the question of terminology) by noting that his "own applications of fiducial probability had been severely and deliberately limited. He had hoped, indeed, that the ingenuity of later writers would find means of extending its application to cases about which he was still in doubt, but some limitations seemed to be essential" (p. 617).[27]

Fisher took it as a logical requirement of an inductive inference that it utilize all available information (here in the form of sufficient statistics or, lacking that, ancillaries), that probability statements not so based are necessarily deficient and that the multiplicity of possible interval estimates that could arise from Neyman's approach was, in effect, symptomatic of its failure to fully utilize the information in a sample. The rationale for the restriction to "exhaustive" statistics was thus *logical* rather than *mathematical*; that is, Fisher insisted on it not because it was necessary for the mathematical validity of the derivation, but because he viewed it as essential for the logical cogency of the resulting statement.

Confidence intervals, Fisher thought in contrast, make statements which, although mathematically valid, are of only limited inferential value. That they do indeed have *some* value was conceded by Fisher in a crucial footnote to his discussion:

Naturally, no rigorously demonstrable statements, such as these are, can fail to be true. They can, however, only convey the truth to those who apprehend their exact meaning; in the case of fiducial statements based on inefficient estimates this meaning must include a specification of the process of estimation employed. But this process is known to omit, or suppress, part of the information supplied by the sample. The statements based on inefficient estimates are true, therefore, so long as they are understood not to be the whole truth. Statements based on sufficient estimates are free from this drawback, and may claim a unique validity. (pp. 617–618)[28]

172

In the remainder of his comments, Fisher made it clear that he did not view this problem as a minor one:

Dr. Neyman claimed to have generalized the argument of fiducial probability, and he had every reason to be proud of the line of argument he had developed for its perfect clarity. The generalization was a wide and very handsome one, but it had been erected at considerable expense, and it was perhaps as well to count the cost. (p. 618)

Fisher then went on to list three specific reservations about Neyman's approach:

1. *The statistics employed were not restricted to those which were exhaustive.* Although Fisher had restricted the discussion in his 1930 paper to estimates arising from the method of maximum likelihood, the requirement there as stated is certainly cryptic, and in later years Fisher faulted his exposition for this reason.[29]

In a paper written shortly after, Fisher (1935a) remedied this omission by reviewing the logic of the fiducial argument in the case of a sample of size n from a normal population with mean μ. If s_1 denotes the sample standard deviation, s_2 the mean absolute deviation and

$$t_j = \frac{(\bar{x} - \mu)/\sqrt{n}}{s_j},$$

then, as Fisher noted, both t_1 and t_2 are pivotal quantities, and each can be employed to derive "probability statements" regarding the unknown parameter μ, although in general the "probability distribution for μ obtained [from t_2] would, of course, differ from that obtained [from t_1]."

There is, however, in the light of the theory of estimation, no difficulty in choosing between such inconsistent results, for it has been proved that, whereas s_2 uses only a portion of the information utilised by s_1, on the contrary, s_1 utilises the whole of the information used by s_2, or indeed by any alternative estimate. To use s_2, therefore, in place of s_1 would be logically equivalent to rejecting arbitrarily a portion of the observational data, and basing probability statements upon the remainder as though it had been the whole. (Fisher, 1935a, p. 393)

2. *The extension to discontinuous variates was only possible by replacing an exact statement of fiducial probability by an inequality.* In particular, Fisher noted, "it raised the question whether exact statements of probability were really impossible, and if they were, whether the inequality arrived at was really the closest inequality to be derived by a valid argument from the data."

This clearly posed mathematical question interested Neyman, and his answer (largely negative) was published the next year (Neyman, 1935b). Fisher's own approach, characteristically clever, was unveiled in his 1935 Royal Statistical Society paper: in some cases a discontinuous variate can be transformed into a continuous variate amenable to the fiducial argument (Fisher, 1935a, pp. 51–53).[30] The problem of fiducial inference for discontinuous variates seems to have exercised a perennial fascination for Fisher; his obituary notice for "Student" gave in passing the simultaneous fiducial distribution for the percentiles of a continuous distribution by means of a discontinuous pivot (Fisher, 1939c, pp. 4–6), and he devoted a lengthy section to the problem of discontinuous variates in his book *Statistical Methods and Scientific Inference* (Fisher, 1956, pp. 63–70).[31]

3. *The extension to several unknown parameters.* Here, too, Fisher saw consistency as a major concern, contrasting the case of a single parameter, where "all the inferences might be summarized in a single probability distribution for that parameter, and that, for this reason, all were mutually consistent," with the multiparameter case, where it had not been shown that "any such equivalent frequency distribution could be established."

Neyman seems to have found this last reservation particularly puzzling,[32] but it clarifies Fisher's interest in the fiducial distribution as guaranteeing that the totality of inferential statements arising from the fiducial argument were mutually consistent.

Thus, Fisher's concerns at this stage were relatively straightforward. He insisted, on logical first principles, that the fiducial argument be limited to exhaustive statistics and saw the multiplicity of interval estimates that could arise from Neyman's approach as symptomatic of the failure of his theory to so limit itself.

4.2. The Break with Neyman

Although initially cordial, the relationship between Fisher and Neyman had never been warm, and in 1935, shortly after the above exchange, relations between the two broke down completely.[33] The occasion of the break was Fisher's discussion of Neyman's 1935 *Journal of the Royal Statistical Society* paper (read 28 March), which was sharply critical of Neyman, both in substance and tone.[34] Neyman's paper had itself been critical of some of Fisher's most important work, although the attack was indirect, and towards Fisher himself the tone of the paper is one of almost studied politeness. (This was not true, however, of Neyman's response.) The reasons for the pointedness of

Fisher's attack can only be conjectured, but with it began a quarter-century long feud which dealt in part with fiducial probability, and we thus enter the second phase of Fisher's writings on the subject.[35]

But before going on to consider this phase, it is important to pause briefly and comment on the Fisher-Neyman dispute itself, because of a nearly universal misapprehension about its nature. Consider, for example, Neyman's description of the feud, summarized in his article "Silver Jubilee of My Dispute with Fisher":

The first expressions of disapproval of my work were published by Fisher in 1935. During the intervening quarter of a century Sir Ronald honored my ideas with his incessant attention and a steady flow of printed matter published in many countries on several continents. All these writings, equally uncomplimentary to me and to those with whom I was working, refer to only five early papers, all published between 1933 and 1938. . . .

Unfortunately, from the very start, [my dispute with Fisher] has been marred by Sir Ronald's unique style involving torrents of derogatory remarks. . . .

Because of my admiration for the early work of Fisher, his first expressions of disapproval of my ideas were a somewhat shocking novelty and I did my best to reply and to explain. Later on, the novelty wore off and I found it necessary to reply only when Fisher's disapprovals of me included insults to deceased individuals for whom I felt respect. My last paper in reply to Fisher [appeared in 1956]. . . . Subsequent polemical writings of Fisher, including a book [Fisher, 1956], I left without reply. (Neyman, 1961, pp. 145–146, references omitted)

This undoubtedly reflected the way Neyman viewed the matter in 1961, but the picture it suggests is almost totally erroneous.

In reality, during the first two decades of the Fisher-Neyman dispute, far from "incessant attention," "a steady flow of printed matter" and "torrents of derogatory remarks," *Fisher almost never referred directly to Neyman in print*. For example, in the first ten years after their break (the period 1935–1944), Fisher referred to Neyman only twice in his papers (Fisher, 1935, 1941), and then only briefly.[36] Likewise, in the decade 1945–1954, one can only find two brief comments related to fiducial inference (Fisher, 1945, 1946); two brief asides in the *Collected Papers* (CP 204 and 205) unrelated to fiducial inference (one of which is innocuous) and a derogatory comment in *Contributions to Mathematical Statistics* (Fisher, 1950). In length, these five passages might comprise a total of two pages of text.

The Fisher-Neyman feud, of course, took place: the poisonous atmosphere in the University College Common room that their two groups shared is legendary. But initially it did not take place, for the most part, in print.[37] The one major exception (Neyman, 1941) was an attack on Fisher by Neyman and did not draw a response from Fisher. All this changed with the publication of Fisher's 1955 *Journal of the Royal Statistical Society* paper, and his 1956 book *Statistical Methods and Scientific Inference*, both of which repeatedly and sharply attacked Neyman in often highly uncomplimentary terms. But for the preceding twenty years of their feud, Fisher chose largely to *ignore* Neyman, and it is Fisher's 1955 paper and 1956 book, which Neyman identifies as the point when he, Neyman, withdrew from the fray, that in reality marks when Fisher's attack first began in earnest (for reasons that will be discussed below).

5. MULTIPARAMETER ESTIMATION

Neyman's claim to have gone beyond Fisher by developing methods for treating the case of several parameters must have seemed an obvious challenge. In a paper published soon after, Fisher presented an extension of the fiducial argument providing a solution to the problem of estimating the difference of two means, the so-called *Behrens-Fisher problem* (Fisher, 1935a; see, generally, Wallace, 1980).[38]

5.1. Fisher's 1935 Paper

Although Fisher emphasized in his 1935 paper (1935a) the necessity of using exhaustive estimates, he did not yet argue for the fiducial solution on the grounds of its conditional nature. Indeed, at one point, while comparing the Bayesian and fiducial approaches, Fisher actually stressed the *unconditional* nature of the fiducial argument:

It is of some importance to distinguish [fiducial] probability statements about the value of μ, from those that would be derived by the method of inverse probability.... The inverse probability distribution would specify the frequency with which μ would lie in any assigned range $d\mu$, by an absolute statement, true of the aggregate of cases in which the observed sample yielded the particular statistics $\bar{x}$ and s. The [fiducial distribution] is independent of all prior knowledge of the distribution of μ, *and is true of the aggregate of all samples without selection. It involves $\bar{x}$ and s as parameters, but does not apply to any special selection of these quantities.* (Fisher, 1935a, p. 392, emphasis added)

Thus, Fisher's conditional concerns did not arise from his dispute with Neyman but arose rather, as will be seen, because of his exchange with Bartlett.

Fisher's 1935 paper contains two important innovations that were to have a profound impact on the direction the fiducial debate later took. The first of these was the introduction of the *simultaneous fiducial distribution* (SFD); the second, the application of such distributions to *multiparameter estimation*.

Maurice Bartlett, a young English statistician, soon raised important concerns about both of these innovations, and Bartlett's concerns, in one way or another, were to be at the heart of many of the later criticisms of fiducial inference. Let us consider each in turn.

5.2. The Simultaneous Fiducial Distribution

Fisher began by setting himself the problem of deriving a "unique simultaneous distribution" for the parameters of the normal distribution. The solution he proposed was ingenious. First illustrating how the fiducial argument could be employed, given a sample of size n_1 from a normal population with unknown μ and σ, to find the fiducial distribution of a single further observation (rather than, as before, unknown population parameters), Fisher showed how this approach could be generalized to obtain a fiducial distribution for the sample statistics $\bar{x}$ and s arising from a second sample of size n_2, and then, by letting $n_2 \to \infty$, Fisher obtained a joint distribution for the population parameters μ and σ.

Where Fisher's 1930 paper had been cautious, careful and systematic, his 1935 paper was bold, clever but in many ways rash. For he now went on to conclude:

In general, it appears that if statistics $T_1, T_2, T_3, \ldots$ contain jointly the whole of the information available respecting parameters $\theta_1, \theta_2, \theta_3, \ldots$, and if functions $t_1, t_2, t_3, \ldots$ of the T's and θ's can be found, the simultaneous distribution of which is independent of $\theta_1, \theta_2, \theta_3, \ldots$, then the fiducial distribution of $\theta_1, \theta_2, \theta_3, \ldots$ simultaneously may be found by substitution. (Fisher, 1935a, p. 395)

This sweeping claim illustrates the purely intuitive level at which Fisher was operating in this paper, and it was only toward the very end of his life that Fisher began to express doubts about this position.[39]

Fisher regarded the SFD as an ordinary probability distribution which could be manipulated in the usual ways, noting, for example, that the marginal distributions of the SFD for (μ, σ) were the previously known fiducial distributions for the two separate parameters. It was at this point that Fisher fell into a subtle trap; for in general the distribution of a function $f(\mu, \sigma)$ of the population parameters, induced by the SFD of μ and σ, will not satisfy the confidence property. The phenomenon already occurs, and is most easily understood, at the univariate level. If X has a $N(\mu, 1)$ distribution, then the fiducial distribution for μ given $X = x$ is $N(x, 1)$, in the sense that if $P_\mu[\mu - X < c_\alpha] = \alpha$, then $P_\mu[\mu < X + c_\alpha] = \alpha$. If, however, the parameter of interest is μ^2, the "fiducial distribution" for μ^2 cannot be derived from that of μ in the usual

177

way that the probability distribution for a random variate U^2 can be derived from that of U, if it is required that the limits arising from such a distribution satisfy the coverage property of Fisher's 1930 paper.[40]

This gap in Fisher's reasoning was later noted by Bartlett (1939), who pointed out that in the case of a normal sample the existence of the simultaneous distribution for (μ, σ) did not (for example) "imply that a fiducial inference could be made for ... $\mu + \sigma$ by integration of the ... fiducial distribution" (p. 133) and that, save in the very special case of the marginals of the SFD, "integration in any other problem is so far justified merely by analogy, and no statement as to its meaning in general has been given by Fisher" (p. 135). Bartlett's point here was completely correct, but his choice of example was exceedingly unfortunate, for it turns out that the *only* (!) univariate functions of (μ, σ) for which the confidence property is preserved when the SFD is integrated are precisely the linear functions $a\mu + b\sigma$ (e.g., Pedersen, 1978). Fisher pounced, and immediately pointed out the absence of any difficulty in the $\mu + \sigma$ example suggested by Bartlett (Fisher, 1941, pp. 143–146; see also 1956, pp. 125–127, 169).

Consistency questions such as these were basic to much of the fiducial debate in the 1950s, but at the time the ease with which Fisher answered Bartlett's specific question about the estimation of $\mu + \sigma$ may have seemed convincing enough to many. Bartlett's other objection to Fisher's multiparameter theory was not, however, so easily dealt with.

5.3. The Behrens-Fisher Problem

Fisher illustrated the uses of the simultaneous fiducial distribution with two examples, one of which was the notorious Behrens-Fisher problem. Few could have predicted then that it would generate a debate lasting several decades. Fisher's solution was almost immediately questioned by Bartlett (1936). Bartlett noted that, unlike the examples of the t-statistic, sample standard deviation and correlation coefficient, the interval estimates for $\mu_2 - \mu_1$ advocated by Fisher gave rise to tests with inappropriate levels of significance, in terms of frequencies involving repeated sampling from the same initial population. Although this must have been a rude surprise to Fisher, he quickly replied (Fisher, 1937) – the first in a series of exchanges with Bartlett over the next several years (Bartlett, 1937, 1939; Fisher, 1939a, 1939b, 1941; see also Bartlett, 1965).

Although in these exchanges Fisher professed to see no difficulty, he must in fact have been deeply troubled. It is revealing to read these papers as a group, for while Fisher kept returning to discuss the logic of the test, maintaining in public a confident air that all was well, the *grounds* on which this was asserted were constantly shifting.

Fisher rejected Bartlett's objection, initially (Fisher, 1937), on the not very convincing grounds that it introduced fixed values for the parameters into the argument, which Fisher argued was inconsistent with the assumed fiducial distribution. Fisher cannot have been comfortable with this response to Bartlett, because fixed values for the parameters had of course entered into his own original fiducial argument at one point.[41]

Two years later, when he returned to the question in response to another paper of Bartlett's (Bartlett, 1939; Fisher, 1939b), this defense was silently dropped, and Fisher defended his solution on the much more radical grounds that the very criterion being invoked by Bartlett was irrelevant:

[T]he problem concerns what inferences are legitimate from a unique pair of samples, which supply the data, in the light of the suppositions we entertain about their origin; the legitimacy of such inferences cannot be affected by any supposition as to the origin of other samples which do not appear in the data. Such a population is really extraneous to the discussion. (p. 386)

This marked a major shift in Fisher's position.[42] Contrast, for example, Fisher's statement above with the language in his 1930 and 1935 papers cited earlier or, most strikingly, that in his 1933 paper:

Probability statements of this type are logically entirely distinct from inverse probability statements, and remain true whatever the distribution *a priori* of σ may be. To distinguish them from statements of inverse probability I have called them statements of fiducial probability. This distinction is necessary since the assumption of a given frequency distribution *a priori*, though in practice always precarious, might conceivably be true, in which case we should have two possible probability statements differing numerically, and expressible in a similar verbal form, though necessarily differing in their logical content. The probabilities differ in referring to different populations; that of the fiducial probability is the population of all possible random samples, that of the inverse probability is a group of samples selected to resemble that actually observed." (Fisher, 1933, p. 348)

Fisher's later writings tended to obscure this shift.[43] When Fisher republished a companion paper to the one above (Fisher, 1939a) in his 1950 collection *Contributions to Mathematical Statistics*, he singled out this point for comment in his introductory note:

Pearson and Neyman have laid it down axiomatically that the level of significance of a test must be equated to the frequency of a wrong decision "in repeated samples from the same population." This idea was foreign to the development of tests of significance given by the author in 1925, for the experimenter's experience does not consist in repeated samples from the same population, although in simple cases the numerical values are often the same. . . . It was obvious from the first, and particularly emphasized by the present author, that Behrens' test rejects a smaller proportion of such repeated samples than the proportion specified by the level of significance, for the sufficient reason that the variance ratio of the populations sampled was unknown.

Such a statement is curiously inconsistent with Fisher's own earlier work. (See especially CP 48, pp. 503–505.) There is no hint in Fisher's 1934 contribution to the Neyman-Pearson theory of uniformly most powerful tests (Fisher, 1934) that he then considered their views to be "foreign to the idea of tests of significance," and when Fisher wrote to Neyman in 1932 commenting on the manuscript of the paper by Neyman and Pearson that later appeared in the *Philosophical Transactions* (Neyman and Pearson, 1933), his primary criticism was on a point of mathematical detail.[44] The assertion by Fisher that "the experimenter's experience does not consist in repeated samples from the same population" stands in contrast with the approach taken by him in his earliest papers on fiducial inference, where the argument is clearly cast in those terms. And far from it having been "obvious from the start" that "Behrens' test rejects a smaller proportion of such repeated samples," Fisher had explicitly conjectured in 1937 that this would *not* always be the case for samples of size greater than two.[45]

Fisher, of course, was certainly entitled to change his mind. But if only he had been willing to admit it![46]

In his papers of the 1930s, Fisher was just beginning to grapple with the problems of conditional inference, and his comments on these basic issues are at times brief, fragmentary, even tentative. It is symptomatic of the uncertainty he must have felt at this point that in 1941 he made the extraordinary concession that Jeffreys (1939, 1940), "whose logical standpoint is very different from my own, may be right in proposing that 'Student's' method involves logical reasoning of so novel a type that a new postulate should be introduced to make its deductive basis rigorous" (Fisher, 1941, p. 142).[47]

But when referring to Neyman, no such concession was possible. By 1945, Fisher's view had hardened, and he labeled the criterion that "the level of significance must be equal to the frequency with which the hypothesis is rejected in repeated sampling of any fixed population allowed by hypothesis" as an "intrusive axiom, which is foreign to the reasoning on which the tests of significance were in fact based" (Fisher, 1945, p. 507). Given the earlier frequency statements appearing in his first papers on fiducial inference, this was somewhat disingenuous.[48]

It is of course possible that Fisher's opposition to Neyman's "clarification" was based solely on an inability to accept that someone could improve on what he had already done.[49] But the evidence clearly suggests otherwise. Even in his discussion of Neyman's 1934 paper, Fisher had emphasized the necessity of utilizing all of the information in a sample; this was basic to Fisher's theory of statistical inference, pervasive in his earlier writings and implicit in his 1930 paper. Indeed, Fisher later claimed to have always insisted on it.[50] This was, moreover, precisely the time when Fisher was grappling with the difficulties of conditional inference, and in later exchanges Fisher would increasingly stress the importance in inference of conditioning on all relevant information. This is indeed a problem that the theory of confidence intervals has yet to resolve.

6. THE YEARS 1942–1955

That Fisher was uncomfortable with the theoretical underpinnings of his fiducial theory is suggested by the direction of his work during the next decade and a half: from 1942 to 1954 Fisher wrote almost nothing on fiducial inference, save a brief expository paper in *Sankhyā* (Fisher, 1945), a letter to *Nature* (Fisher, 1946), an expository paper in French (Fisher, 1948) and a discussion of a paper by Monica Creasy (Fisher, 1954).[51]

In his 1945 paper, Fisher illustrated the fiducial argument with the simple example of a sample of two drawn from a continuous distribution having a (presumably unique) median μ.[52] If X denotes the number of observations less than the median, then X is a pivotal quantity with binomial distribution $B(2, 1/2)$; thus, for example, $P[\mu < \min(X_1, X_2)] = 1/4$. Fisher reasoned: "recognizing this property we may argue from two given observations, now regarded as fixed parameters that the probability is $1/4$ that μ is less than both x_1 and x_2 .... The idea that probability statements about unknown parameters cannot be derived from *data* consisting of observations can only be upheld by those willing to reject this simple argument" (Fisher, 1945, p. 131).

This is candid enough, but it is really a complete admission of failure: it was precisely the cogency of this "simple argument" that Neyman (1941) and others had so vocally questioned.[53] Fisher could no longer appeal to the unconditional sampling justification of his earliest papers, but he was unable to supply an alternative given his new, conditional view of the matter. It was *intuitively* obvious to Fisher that the existence of a pivot warranted the transition from a probability assertion about statistics (conditional on the parameter) to a probability assertion about parameters (conditional on the statistic), but the significance of the passage quoted is that its language reveals that at this point Fisher was totally unable to supply further support or justification for that intuition.

This state of affairs lasted for a decade. Fisher's 1955 attack on the Neyman-Pearson approach to statistical inference in the *Journal of the Royal Statistical Society* (Fisher, 1955) touched only briefly on the specific question of fiducial inference, but, brief as his comments there are, they make it abundantly clear that he was no nearer to a satisfactory justification for the logical inversion central to the fiducial argument than he had been ten years earlier, when he wrote his expository piece for *Sankhyā*:

A complementary doctrine of Neyman violating equally the principles of deductive logic is to accept a general symbolical statement such as

$$Pr\{(\bar{x} - ts) < \mu < (\bar{x} + ts)\} = \alpha,$$

as rigorously demonstrated, and yet, when numerical values are available for the statistics $\bar{x}$ and s, so that on substitution of these and use of the 5 per cent. value of t, the statement would read

$$Pr\{92.99 < \mu < 93.01\} = 95 \text{ per cent.},$$

to *deny* to this *numerical* statement any validity. This is to deny the syllogistic process of making a substitution in the major premise of terms which the minor premise establishes as equivalent. (p. 75)

A year later, however, in 1956, Fisher felt he had finally achieved a coherent rationale for the fiducial argument.

7. THE LAST BATTLE

Fisher's treatment of the fiducial argument in *Statistical Methods and Scientific Inference* (cited within as *SMSI*) (and nearly a dozen papers during the next several years) marks the third and final phase in his advocacy of fiducial inference. Perhaps realizing that he was now fighting a clearly downhill battle, Fisher made an obvious effort to present a clear statement of its logic.[54] Indeed, he went so far as to concede that "the applicability of the probability distribution [of the pivotal statistic] to the particular unknown value of [the parameter] . . . on the basis of the particular value of T given by his experiment, has been disputed, and certainly deserves to be examined" (*SMSI*, p. 57). This was a remarkable admission, since only a year earlier Fisher had excoriated Neyman for questioning precisely this applicability (Fisher, 1955, pp. 74–75)![55]

But in the interim Fisher's view of fiducial inference had radically altered. As he himself described it:

It is essential to introduce the absence of knowledge *a priori* as a distinctive datum in order to demonstrate completely the applicability of the fiducial method of reasoning to the particular real and experimental cases for which it was developed. This point I failed to perceive when, in 1930, I first put forward the fiducial argument for calculating probabilities. For a time this led me to think that there was a difference in logical content between probability statements derived by different methods of reasoning. There are in reality no grounds for any such distinction. (*SMSI*, p. 59)

One might assume from Fisher's wording ("for a time this led me to think") that this shift in his thinking had occurred many years earlier. But that it had occurred only a short time earlier is evident from a passage in the *Design of Experiments* (1935c; see also 1951, p. 42). As late as the 6th revised edition of 1953 Fisher had continued to assert (emphasis added) the following:

Statements of inverse probability have a different logical *content* from statements of fiducial probability, in spite of their similarity of form, *and* they require for their truth the postulation of knowledge beyond that obtained by direct observation. (Section 63)

But by 1956 Fisher no longer believed this, and thus in the next edition (7th, 1960) of *The Design of Experiments*, published a few years later, Fisher changed "content" to "basis" (as well as "and" to "for"). A basic and fundamental shift in Fisher's view of the nature of fiducial inference has thus been silently disguised by the subtle change of a single word. When Fisher wrote that inverse and fiducial statements differ in their *content*, he was referring primarily (at least in 1935, when this passage was first written) to the conditional aspect of the former and the unconditional aspect of the latter. But when he says that they differ in their logical *basis*, he intends something quite different: both are, to use Dempster's phrase, "postdictive," but in one case based on prior knowledge (that is, a postulated prior distribution for the unknown parameter), in the other on the *absence* of prior knowledge of the parameter. But just exactly what does the verbal formulation "absence of prior knowledge" mean? Fisher had very early on rejected the Bayesian move that attempted to translate this into a uniform prior distribution for the parameter, for, as he noted (Fisher, 1922, p. 325), uniformity of prior is not invariant under parametric transformation. His insight in 1955–1956 was that the verbal, qualitative formulation "absence of prior knowledge" could be translated into an exact, quantitative postdictive distribution by invoking the fiducial argument – that the "absence of prior knowledge" was *precisely* the epistemological state which justified the invocation of the fiducial argument.

This shift reflects in part his new view of the nature of probability and in part the device of recognizable subsets.

7.1. The Nature of Probability (continued)

Fisher's treatment of probability in *SMSI* reveals an apparent shift in his view of its nature. In his papers before World War II, Fisher had described prior distributions as referring to an objective process by which population parameters were generated. For example, writing in 1921, Fisher states that the problem of finding a posterior distribution "is indeterminate without knowing the statistical mechanism under which different values of [a parameter] come into existence" (Fisher, 1921, p. 24) and that "we can know nothing of the probability of hypotheses or hypothetical quantities" (p. 35). (In the 1950 introduction to this paper in *Contributions to Mathematical Statistics* (Fisher, 1950), Fisher brands the second assertion as "hasty and erroneous.")

In contrast, in the 1950s Fisher espoused a view of probability much closer to the personalist or subjectivistic one: "probability statements do not imply the existence of [the hypothetical] population in the real world. All that they assert is that the exact nature and degree of our uncertainty is just *as if* we knew [the sample] to have been one chosen at random from such a population" (Fisher, 1959, p. 22). None of the populations used to determine probability levels in tests of significance have "objective reality, all being products of the statistician's imagination" (Fisher, 1955, p. 71; cf. *SMSI*, p. 81). In the first and second editions of *SMSI*, Fisher referred to "the role of subjective ignorance, as well as that of objective knowledge in a typical probability statement" (p. 33). [This embrace of the subjective was apparently too radical, however, for someone who had once tagged Jeffreys's system as "subjective and psychological" (CP 109, p. 3), and in the third edition of *SMSI*, the passage was silently emended to read "the role both of well specified ignorance and of specific knowledge in a typical probability statement" (p. 35).]

Although Fisher remained publicly anti-Bayesian, after World War II he was in fact much closer to the "objective Bayesian" position than that of the frequentist Neyman.[56] In a little noted passage in *SMSI*, Fisher even cited without criticism Sir Harold Jeffreys's Bayesian derivation of the Behrens-Fisher interval, saying only that Jeffreys and others, recognizing "the rational cogency of the fiducial form of argument, and the difficulty of rendering it coherent with the customary forms of statement used in mathematical probability," had introduced "new axioms to bridge what was felt to be a gap," whereas "[t]he treatment in this book involves no new axiom" (p. 59). This was somewhat remarkable, inasmuch as Jeffreys's new axioms were modern reformulations of Bayes's postulate!

7.2. Recognizable Subsets

In Fisher's new view, an assertion of probability contained three elements: the specification of a reference set, the assertion that the outcome of interest was an element of this set and the assertion that no subset of the reference set was "recognizable" (*SMSI*, p. 60).[57] In the case of estimation, Fisher thought the absence of recognizable subsets a consequence of the requirements that the statistics employed be exhaustive and that there be absence of prior knowledge regarding the parameters (*SMSI*, p. 58). As an illustration, Fisher cited the case of the *t*-statistic:

[T]he inequality

$$\mu < \bar{x} - \frac{1}{\sqrt{N}} ts$$

will be satisfied with just half the probability for which *t* is tabulated, if *t* is positive, and with the complement of this value if *t* is negative. The reference set for which this probability statement holds is that of the values of μ, $\bar{x}$ and s corresponding to the same sample, for all samples of a given size of all normal populations. Since $\bar{x}$ and s are jointly Sufficient for estimation, and knowledge of μ and σ *a priori* is absent, there is no possibility of recognizing any sub-set of cases, within the general set, for which any different value of the probability should hold. (*SMSI*, p. 84; cf. Fisher, 1959, pp. 25–26)

This was clear enough, but Fisher's assertion, that the use of exhaustive estimates and the lack of knowledge a priori combined to insure the absence of recognizable subsets, was just that, an assertion. Seven years later Buehler and Feddersen (1963) somewhat unexpectedly showed that in precisely this case of the *t*-distribution recognizable subsets *did* exist, thus decisively refuting Fisher's final and clearest attempt at a justification.

A letter from Fisher to Barnard written in 1955 (14 October, Bennett, 1990, pp. 31–32) is revealing. Barnard, who had read a draft of this chapter in *SMSI*, queried Fisher about the justification for the fiducial distribution of μ, and whether it was not based on the joint distribution of μ and σ. In reply, Fisher wrote that he did not think so, arguing as above:

[I]f it is admitted that no subset can be recognized having a different probability, and to which the observed sample certainly belongs, (as can scarcely be disputed since $\bar{x}$

and *s* are jointly sufficient and it is postulated that no information *a priori* is available), the distribution of μ follows from that of *t*. (p. 32)

It is clear from Fisher's wording ("as can scarcely be disputed") that the basis for this assertion was an intuitive conviction on Fisher's part rather than a mathematical demonstration, and in a subsequent letter (on 17 October) a sceptical Barnard continued to query the point.

8. AFTERMATH

Although fiducial inference had its advocates in the years 1935–1955, a substantial majority of the statistical profession preferred the conceptual clarity of Neyman's confidence interval approach, and relatively few papers appeared on fiducial inference during this period. All this changed with the appearance of Fisher's book, which sparked renewed interest and controversy.

But that debate was largely possible only because of the ambiguities inherent in Fisher's theory (especially the method by which simultaneous fiducial distributions were to be constructed), his willingness in many instances to rely on intuition when asserting matters of mathematical fact and his preference for basing his treatment on "the semantics of the word 'probability'" (*SMSI*, p. 59), rather than axiomatics.[58]

Only in correspondence did Fisher express uncertainties never voiced publicly. As Fisher's friend George Barnard (1963) has noted, Fisher's "public utterances conveyed at times a magisterial air which was far from representing his true state of mind [regarding the fiducial argument in the case of several parameters]. In one letter he expresses himself as 'not clear in the head' about a given topic, while in another he referred ruefully to 'the asymptotic approach to intelligibility'" (p. 165). Indeed, Fisher once confessed to Savage, "I don't understand yet what fiducial probability does. We shall have to live with it a long time before we know what it's doing for us. But it should not be ignored just because we don't yet have a clear interpretation" (Savage, 1964, p. 926; see also Box, 1978, p. 458).

Once Fisher had gone from the scene, much of the heart went out of the fiducial debate, although important contributions continued to be made, most notably in the structural approach of Fraser and the belief function approach of Dempster. This literature was concerned not so much with fiducial inference, in the form Fisher conceived it, but with the attempt to achieve the goals for which it had been initially, if unsuccessfully, forged. As such it is beyond the scope of this paper. Three important papers which provide an entry into much of this literature are those of Wilkinson (1977), Pedersen (1978) and Wallace (1980).

186

The fiducial argument stands as Fisher's one great failure. Not only did he stubbornly insist on its cogency, clarity and correctness long after it became clear that he was unable to provide an understandable statement of it, let alone a coherent theory (Savage, 1976, p. 466, refers to Fisher's "dogged blindness about it all"), but he later engaged in a futile and unproductive battle with Neyman which had a largely destructive effect on the statistical profession. In *SMSI*, he was candid enough to confess the inadequacy of his earlier attempts to describe the fiducial argument and indiscreet enough to restate the argument with a clarity which permitted it to be decisively refuted.

Before his dispute with Neyman, Fisher had engaged in other statistical controversies, crossing swords with Arthur Eddington, Harold Jeffreys and Karl Pearson.[59] He had been fortunate in his previous choice of opponents: Eddington conceded Fisher's point, Jeffreys was cordial in rebuttal and Pearson labored under the disadvantage of being completely wrong.

But, in Neyman, Fisher was to face an opponent of an entirely different character.

9. CONCLUSION

The fiducial argument arose out of Fisher's desire to create an inferential alternative to inverse methods, avoiding the arbitrary postulates on which the classical Laplacean approach depended. Fisher felt he had discovered such an alternative in 1930, when he realized that the existence of pivotal quantities permitted the derivation of a probability distribution for an unknown parameter "irrespective of any assumption as to its *a priori* distribution" (p. 533).

The original fiducial argument, for a single parameter, was virtually indistinguishable from the confidence approach of Neyman, although Fisher thought its application should be restricted in ways that reflected his view of the logical basis of inductive reasoning. This effectively blended both an inferential and a behaviorist viewpoint. When Fisher subsequently attempted to extend the fiducial argument to the multiparameter setting in his treatment of the Behrens-Fisher problem, this conflict surfaced, and, forced to decide between the two, Fisher opted for the inferential, rather than the behaviorist route, thus (silently) abandoning the unconditional sampling approach of his earlier papers for the conditional approach of his later work.

Initially unable to justify his intuition about the passage from a probability assertion about a statistic (conditional on a parameter) to a probability assertion about a parameter (conditional on a statistic), Fisher thought in 1956 that he had finally discovered the way out of this enigma with his concept of

187

recognizable subset. But despite the authoritative way in which Fisher asserted his new position in his last book, *Statistical Methods and Scientific Inference*, the crucial argument for the relevance of this concept was founded on yet another intuition – one which, now clearly stated, was later demonstrated to be false by Buehler and Feddersen in 1963.

Fiducial inference in its final phase was in essence an attempt to construct a theory of conditional confidence intervals (although Fisher would never have put it that way) and thereby "make the Bayesian omelette without breaking the Bayesian eggs." Fisher's failure, viewed in this light, was hardly surprising: no satisfactory theory of this type yet exists. But Fisher's attempt to steer a path between the Scylla of unconditional, behaviorist methods which disavow any attempt at "inference" and the Charybdis of subjectivism in science was founded on important concerns, and his personal failure to arrive at a satisfactory solution to the problem means only that the problem remains unsolved, not that it does not exist.

ACKNOWLEDGMENT

At various stages in the drafting of this paper, I have received valuable and generous assistance and comments from many people interested in Fisher and his work. These include George Barnard, Maurice Bartlett, Arthur Dempster, Anthony Edwards, Erich Lehmann, Paul Meier, Teddy Seidenfeld and David Wallace. I am very grateful to them all.

NOTES

1. See his statement in Fisher (1956, p. 77). In CP 290, Fisher contrasts the results of the fiducial argument with "those weaker levels of uncertainty represented by Mathematical Likelihood, or only by tests of significance."
2. Soper et al. (1917). After erroneously stating that "[Fisher] holds that *a priori* all values of ρ are equally likely to occur" (p. 353), the authors discussed the consequences of assuming instead a Gaussian prior. For Fisher's reply, see Fisher (1921); see also Pearson (1968, pp. 452–454), Edwards (1974) and Box (1978, p. 79).
3. Egon Pearson (1968) gives the text of the letter from Pearson to Fisher rejecting the 1921 paper. In addition to the two papers just mentioned, Pearson had also earlier rejected a note by Fisher briefly criticizing an article in the May 1916 issue of *Biometrika*; see Pearson (1968, pp. 454–456). See also Pearson (1990).
4. "[H]e [Pearson] gained the devoted service of a number of able assistants, some of whom he did not treat particularly well. He was prolific in magnificent, or grandiose, schemes capable of realization perhaps by an army of industrious robots responsive to a magic wand. . . . The terrible weakness of his mathematical and scientific work flowed from his incapacity in self-criticism, and his unwillingness to admit the

possibility that he had anything to learn from others, even in biology, of which he knew very little. His mathematics, consequently, though always vigorous, were usually clumsy, and often misleading. In controversy, to which he was much addicted, he constantly showed himself to be without a sense of justice. His immense personal output of writings . . . left an impressive literature. The biological world, for the most part, ignored it, for it was indeed both pretentious and erratic" (Fisher, 1956, pp. 2–3). As Savage has noted, Fisher "sometimes published insults that only a saint could entirely forgive" (Savage, 1976, p. 446). See also Kendall (1963, p. 3) and Barnard (1990, p. 26).

5. Karl Pearson's Laplacean view of probability is most carefully set out in Chapter 4 of his *Grammar of Science* (1892); see also Pearson (1920). Although little read today, the impact of Pearson's *Grammar* in his own time was considerable. (Mach's *Science of Mechanics*, for example, was dedicated to Pearson.) For the influence of the *Grammar* on Neyman, see Reid (1982, pp. 24–25).

6. I owe the material in this section to the generosity of Dr. A. W. F. Edwards, who has made available to me a considerable body of information he collected during the 1970s about the Rothamsted origins of the fiducial argument.

7. Fisher (1935c), in the chapter entitled "The Generalisation of Null Hypotheses. Fiducial Probability."

8. Because Fisher later distanced himself so emphatically from Neyman's viewpoint, the development of the theory of confidence intervals eventually came to be associated almost exclusively with Neyman (and his school). But, although Fisher disagreed with Neyman's behaviorist interpretation and unconditional uses of confidence intervals, Fisher's priority in the discovery of the method itself – in terms of publication relative to Neyman – seems largely unappreciated. [*Relative* rather than absolute priority: "E. L. Lehmann has pointed out that as far as computation (as opposed to logic) is concerned there is a long tradition of constructing confidence intervals involving Laplace and Poisson, followed by Lexis and one may add Cournot" (Hacking, 1990, p. 210). The reference is to a 1957 technical report written by Lehmann; in a footnote, Hacking notes that "Lehmann's paper has never been published, originally because he did not wish to offend Neyman" and cites as his source a personal letter from Lehmann dated 5 July 1988.]

9. Fisher does not specify with exactitude the necessary conditions on F, and his notation has the disadvantage that it does not distinguish between the random variate T and an observed value of that variate.

10. Using this time the example of estimating σ from s, on the basis of a random sample from a normal population, Fisher wrote:

Now we know that the inequality $[s > s_{0.01}(\sigma)]$ will be satisfied in just 1 per cent. of random trials, whence we may infer that the inequality $[\sigma < \sigma_{0.99}(s)]$ will also be satisfied with the same frequency. Now this is a probability statement about the unknown parameter σ. (Fisher, 1933, pp. 347–348)

11. This is underscored in a letter of Fisher to Fréchet several years later (26 February 1940; Bennett, 1990), where such a frequency is said to be "a definition of the phrase fiducial probability," one that Fisher had "no objection to regarding . . . as an arbitrary definition" (p. 127).

12. Fisher termed a probability resulting from the fiducial argument a "fiducial probability," to distinguish it from an "inverse probability" (Fisher, 1933, p. 83; 1945, p. 129), but stressed that while the terminology was intended to draw attention to the novel mode of derivation employed, such probabilities did not differ in kind from ordinary mathematical probabilities (Fisher, 1936) and that "the concept of probability involved is entirely identical with the classical probability of the early writers, such as Bayes" (Fisher, 1956, p. 54).

   For the distinction between probability and likelihood, see Fisher (1921, pp. 24–25; 1922, pp. 326–327).

13. "For a time this led me to think that there was a difference in logical content between probability statements derived by different methods of reasoning. There are in reality no grounds for any such distinction" (Fisher, 1956, p. 59).

14. For example, Wilks (1938).

15. As Fisher (1939b) later put it: "To all, I imagine, it [the fiducial argument] implies at least a valid test of significance expressible in terms of an unknown parameter, and capable of distinguishing, therefore those values for which the test is significant, from those for which it is not" (p. 384). See also Fisher (1935b, pp. 50–51).

16. "[Fiducial inferences] are certainly not statements of the distribution of a parameter θ over its possible values in a population defined by random samples selected to give a fixed estimate T" (Bennett, 1990, p. 124).

17. The fiducial distribution is the "*aggregate* of all such statements as that made above" (Fisher, 1936, p. 253); "une loi de probabilité pour μ qui correspondra à l'ensemble des résultats trouvés plus haut" (CP 156, p. 155).

18. For example, in a letter in 1934 to Harold Jeffreys Fisher considered the amusing example of a traveller landing by parachute in a city and finding that he is one kilometer from its center. If the city is assumed circular and the position of the traveler random, then the fiducial probability that the radius of the city exceeds R kilometers is $1/R^2$. Thus, the "fiducial median city has a radius $\sqrt{2}$ kilometres and an area 2π. The fiducial mean radius is 2 km. and the fiducial mean area is infinite" (Bennett, 1990, pp. 160–161).

19. In particular, Fisher (1922, 1925, 1930, 1934, 1935b, 1936).

20. Fisher (1934, pp. 6–7; 1935b, p. 40).

21. Bennett (1990, p. 61).

22. Fisher (1934, p. 4).

23. Fisher (1922, p. 326); see also Bennett (1990, pp. 172–173). Fisher considers the theories of Ellis and Cournot to be "sound" (Bennett, 1990, p. 61).

24. Fisher (1925, p. 700; 1934, p. 7). "I myself feel no difficulty about the ratio of two quantities, both of which increase without limit, tending to a finite value, and think personally that this limiting ratio may be properly spoken of as the ratio of two infinite values when their mode of tending to infinity has been properly defined" (Bennett, 1990, p. 151; see also pp. 172–173).

25. CP 109, p. 7; CP 124, p. 81. Fisher thought that failure to maintain a clear distinction between the hypothetical and experimental value of probability was responsible for the lack of universal acceptance of the frequency theory (Bennett, 1990, p. 61).

26. This discussion is unfortunately omitted from Fisher's *Collected Papers*. All quotations in this section, unless otherwise stated, are from the report of the discussion at the end of Neyman (1934).

27. Strictly speaking, it would have been more accurate here for Fisher to have referred to the fiducial *argument*, rather than fiducial *probability*.
28. Note the use of the expression "fiducial statements based on inefficient estimates"; the fiducial argument may be employed in such cases, although care is needed in the use and interpretation of the resulting intervals. Fisher did not abandon this stand after his break with Neyman. In his 1956 book, Fisher wrote that "[confidence limits,] though they fall short in logical content of the limits found by the fiducial argument, and with which they have often been confused, do fulfil some of the desiderata of statistical inferences" (p. 69).
29. Fisher (1935a). In his introduction to the 1930 paper in *Contributions to Mathematical Statistics* (Fisher, 1950), Fisher states that "it should also have been emphasised that the information [supplied by a statistic employed in a statement of fiducial probability] as to the unknown parameter should be exhaustive" (pp. 392–393), and Fisher (1956) states that "though the [correlation coefficient] example was appropriate, my explanation left a good deal to be desired" (p. 57). See also Bennett (1990, pp. 81–82).
30. During the discussion, Neyman complimented Fisher on using a "remarkable device" to alter the problem "in such an ingenious way" (Fisher, 1935b, p. 76).
31. One reason for this particular interest might have been Fisher's often expressed view that the fiducial argument had not been noted earlier because of the "preoccupation" of earlier authors with discontinuous variates, to which the particular argument given in his 1930 paper did not apply (see, e.g., Fisher, 1935a, p. 391; 1941, p. 323).
32. Neyman later wrote, "Fisher took part in the discussion, and it was a great surprise to the author to find that, far from recognizing them as misunderstandings, [Fisher] considered fiducial probability and fiducial distributions as absolutely essential parts of his theory" (Neyman, 1941, p. 129). Although one cannot be sure how closely the published text of Fisher's remarks mirrors his actual words, Neyman's statement is certainly not supported by the published version of the discussion. Far from asserting that fiducial probabilities were a novel element of his theory, Fisher agreed with Neyman that they did not differ from ordinary probabilities, the adjective "fiducial" only being used to indicate "a probability inferred by the fiducial method of reasoning, then unfamiliar, and not by the classical method of *inverse* probability." The term "fiducial distribution" itself does not appear in Fisher's discussion. (Fisher did state that "with a single parameter, it could be shown that all the inferences might be summarized in a single probability distribution for that parameter, and that, for this reason, all were mutually consistent; but it had not yet been shown that when the parameters were more than one any such equivalent frequency distribution could be established.") Neyman would appear to be largely projecting back to 1935 ideas and statements made only later by Fisher.
33. It is interesting to contrast the treatment of this period in Constance Reid's biography of Neyman (Reid, 1982) and Joan Fisher Box's biography of Fisher (Box, 1978). Reid, who interviewed both Neyman and Pearson, paints a picture of continuing cordial relations until March 1935: "Jerzy – to start with – got on quite well with Fisher" (p. 114, quoting Egon Pearson); throughout the spring of 1934, "Neyman continued to be on good terms with Fisher; and he was invited, as he recalls, several times to Rothamsted" (p. 116); the next fall, "Neyman continued to receive friendly invitations to Rothamsted" (p. 120); and in December Neyman's "highly

complimentary" remarks on Fisher's RSS paper (read on 18 December 1934) "drew grateful words from the beleaguered Fisher" (p. 121). In contrast, Box writes that after Fisher nominated Neyman for membership in the ISI in May 1934, "Neyman sniped at Fisher in his lectures and blew on the unquenched sparks of misunderstanding between the departments [of Genetics and Statistics at University College London] with apparent, if undeliberate, genius for making mischief," resulting in "open conflict" after the reading of Neyman's 1935 paper, whose "condescending attitude would have been galling, even if the conclusion had been sound" (p. 263).

34. "Professor R. A. Fisher, in opening the discussion, said he had hoped that Dr. Neyman's paper would be on a subject with which the author was fully acquainted, and on which he could speak with authority, as in the case of his address to the Society delivered last summer. Since seeing the paper, he had come to the conclusion that Dr. Neyman had been somewhat unwise in his choice of topics." First describing a statement in Neyman's paper as "extraordinary," Fisher later asked "how had Dr. Neyman been led by his symbolism to deceive himself on so simple a question?" and ended by referring to "the series of misunderstandings which [Neyman's] paper revealed" (Neyman, 1935a, pp. 154–157. Fisher's posthumously published *Collected Papers* presents a highly sanitized version of these comments.)

35. During the next several years, Fisher would provoke a series of needless professional and personal confrontations besides that with Neyman: lashing out at this old friend Gosset a year before the latter's death, exacerbating a long controversy in population genetics with Sewall Wright, and worsening relations with his wife, which led to permanent separation. Fisher had moved in 1933 from the congenial atmosphere of Rothamsted to a divisive University College London, and it is possible that in this perhaps unfortunate change lies much of the explanation; another important factor may have been the decidedly unfriendly reception given to his 1935 Royal Statistical Society paper.

36. Although Neyman's comments largely suggest a personal attack, the wording "my ideas" and "those with whom I was working" might also be taken to include attacks on Neyman's work not directly naming him, and attacks on others, not necessarily coauthors. I have not found many instances of the former, however, and the disputes with Bartlett (discussed below), Wilson and Barnard do not appear to fall into the category delineated by Neyman. Who Neyman might have had in mind is unclear.

37. This statement refers only to direct exchanges between the two, and not to others who may have served as proxies; see, for example, the paper by Yates (1939) and its discussion by Neyman (1941).

38. The paper testifies to the sudden deterioration in relations between Fisher and Neyman. Where just a few months earlier Fisher had referred approvingly to Neyman's 1934 paper (see note above), now Fisher wrote, "Dr. J. Neyman has unfortunately attempted to develop the argument of fiducial probability in a way which ignores the results from the theory of estimation, in the light of which it was originally put forward. His proofs, therefore, purport to establish the validity of a host of probability statements many of which are mutually inconsistent" (Fisher, 1935a, p. 319).

39. The statement being made is in fact quite strong. The phrase "it appears that" does not intend the qualified assertion "it *would seem* that" but the unqualified assertion "it *is seen* that"; compare Fisher's use of the expression in the preceding paragraph.

When the paper was reprinted (in his *Collected Papers*), Fisher had added the curious footnote, "After appears, insert likely"!

40. If, given α and x, $\alpha_1(x)$ and $\alpha_2(x)$ denote the unique numbers satisfying the dual constraints $\alpha_1(x) - \alpha_2(x) = \alpha$ and $x + c_{\alpha_2}(x) = -(x + c_{\alpha_1}(x))$, then it is *not* the case that $P_\mu[X + c_{\alpha_2}(X) < \mu < X + c_{\alpha_1}(X)] = \alpha$ (see, e.g., Pedersen, 1978).

41. "[One source of paradoxes] is the introduction, into an argument of this type, of fixed values for the parameters, an introduction which is bound to conflict with the fiducial distributions derivable from the data" (Fisher, 1937, p. 370). Why this would conflict with the fiducial distribution Fisher did not state. In later papers, Fisher attempted to deal with this difficulty by dogmatic decree. Thus, "The notion of repeated sampling from a fixed population has completed its usefulness when the simultaneous distribution of t_1 and t_2 has been obtained" (Fisher, 1941, p. 148).

42. In his introduction to this paper in *Contributions to Mathematical Statistics* (Fisher, 1950), Fisher admitted as much when he pointed to "the first section, in which the logic of the [Behrens-Fisher] test is discussed" and noted that "the principles *brought to light* seem essential to the theory of tests of significance in general" [emphasis added]. But given that those principles had been "brought to light" in 1939, his charge – in the very same sentence – that they had "been most unwarrantedly ignored" by Neyman and Pearson in a paper written *seven years earlier* is curious to say the least.

   An important factor contributing to this shift may have been Fisher's rereading of Gosset's papers while drafting an obituary notice for "Student" (Fisher, 1939c). Fisher's method, first given there, for estimating the median or other percentiles of a distribution "irrespective of the form of curve" (pp. 4–5) stands in striking contrast to his earlier criticism of a paper by Harold Jeffreys (Jeffreys, 1932; Fisher, 1933). Jeffreys had asserted that given two observations x_1 and x_2, the probability is $1/3$ that a third observation x_3 will lie between the first two; Fisher now asserted (p. 4) that given two observations x_1 and x_2, the probability is $1/2$ that the median lies between them. But the repeated sampling argument that Fisher had employed to ridicule Jeffrey's statement in 1933 could be easily modified to attack his own assertion in 1939. (Note also that Fisher's new emphasis on the uniqueness of the sample at hand was also justified by pointing to the wording used earlier by Student; see Fisher, 1939a, p. 175).

   Another important factor contributing to Fisher's shift in viewpoint was undoubtedly his lengthy exchange of letters with Harold Jeffreys between 1937 and 1942 (Bennett, 1990, pp. 161–178). See especially Fisher's comment that "I have just reread your note on the Behrens-Fisher formula. . . . I think your paper enables me to appreciate your point of view [i.e., conditional and Bayesian] a great deal better than I have previously done" (pp. 175–176).

43. Indeed, initially Fisher does not seem to have recognized the inconsistency of the two positions: in a letter to Fréchet in 1940 (26 January; Bennett, 1990, p. 121) Fisher reiterated the position of his 1930 paper that in a statement of fiducial probability the statistics involved are not considered as fixed and that such a statement differs in logical content from one of inverse probability. (See also Fisher's letter to Fréchet dated 10 February 1940; Bennett, 1990, p. 124.) The clash with the language in a paper of only five years later (Fisher, 1945, quoted below in endnote 48) is particularly striking.

44. Fisher pointed out that a distribution is not determined, as had been claimed, by its moments; see Reid (1982, p. 103). Ironically, Fisher may have served as one of the referees for the paper when it was submitted to the *Philosophical Transactions*, reporting favorably on it; see Reid (1982, pp. 102–104).

45. "With samples of more than 2, I should expect some differences fiducially significant to be found insignificant, if tested for some particular values of the variance ratio, these being ratios which the data themselves had shown to be unlikely" (Fisher, 1937, p. 375). In his 1937 paper, Fisher mathematically demonstrated the conservative nature of the Behrens-Fisher test for samples of size two, presumably to verify the universal validity of the phenomenon noted by Bartlett in several specific instances. Although numerical studies suggest that the Behrens-Fisher test is indeed conservative for all sample sizes, as Fisher later asserted in 1950, a mathematical demonstration of this fact is still lacking today (see, e.g., Wallace, 1980, p. 137)!

46. In some cases, of course, Fisher may simply have come to believe in a new position with such force and conviction that he simply forgot that there had ever been a time when he thought otherwise. For example, in a letter to Barnard in 1954 Fisher criticized Neyman for ignoring "my warning [in Fisher, 1930] that the fiducial distribution would be invalid to any one possessing knowledge *a priori* in addition to the observed sample" (Bennett, 1990, pp. 9–10). In reality, as we have seen, far from having issued such a warning, Fisher clearly takes the opposite position!

47. Note the contrast with Fisher's statement in 1935 that "to attempt to define a prior distribution of μ which shall make the inverse statements coincide numerically with the fiducial statements is really to slur over this distinction between the meaning of statements of these two kinds" (Fisher, 1935a, p. 392).

48. This paper, although it does not refer to Neyman (1941), was clearly intended as a reply to it. ["The purpose of this note is therefore to discuss . . . the process of reasoning by which we may pass, without arbitrariness or ambiguity, from forms of statement in which observations are regarded as random variables, having distribution functions involving certain fixed but unknown parameters, to forms of statement in which the observations constitute fixed data, and frequency distributions are found for the unknown parameters regarded as random variables" (Fisher, 1945, p. 507).]

49. See, for example, the rather jaundiced view of Raymond Birge, the Berkeley physicist, quoted in Reid (1982, p. 144). Savage's assessment was much more sympathetic (and probably more accurate): "I am surely not alone in having suspected that some of Fisher's major views were adopted simply to avoid agreeing with his opponents. One of the most valuable lessons of my rereading is the conclusion that while conflict may have sometimes somewhat distorted Fisher's presentation of his views, the views themselves display a steady and coherent development" (Savage, 1976, p. 446, references omitted).

50. "From the time I first introduced the work, I have used the term fiducial probability rather strictly, in accordance with the basic ideas of the theory of estimation. Several other writers have preferred to use it in a wider application, without the reservations which I think are appropriate" (Fisher, 1939b, p. 384).

51. Creasy's paper (1954) dealt with the problem of assigning fiducial limits to a ratio of normally distributed means (the so-called Fieller-Creasy problem); see Wallace (1980). It is generally agreed, even by many of Fisher's most sympathetic readers,

that he was unfair in his critical response to Creasy's paper; see, for example, Box (1978, p. 459) and Wallace (1980, p. 141).

52. The more general case of estimating the percentiles of a distribution on the basis of a sample of size n had been discussed earlier by Fisher in his obituary of 'Student' (Fisher, 1939c).

53. Fisher's correspondence with Fréchet in 1940 (Bennett, 1990, pp. 118–134) is particularly interesting in this regard. Repeatedly pressed by Fréchet to justify the transition, Fisher eventually argued that the particular sample (and therefore the resulting interval) could be regarded as "an event drawn at random from the population investigated" and therefore that the single-case probability of coverage could be identified with the frequency of coverage in the population as a whole.

54. See Fisher's remarks in the preface to the 13th edition of *Statistical Methods for Research Workers* (Fisher, 1925, 13th ed, 1958).

55. The language of this passage suggests that, as late as the beginning of 1955, Fisher had not yet arrived at his recognizable subset justification for the fiducial argument.

56. See, for example, Box (1978, pp. 441–442). I. J. Good reports that he had been told Fisher liked his 1950 book *Probability and the Weighing of Evidence* (Savage, 1976, p. 492; see also Bennett, 1990, p. 137). As Barnard notes, Jeffreys "got on extremely well with Fisher" (Barnard, 1990, p. 27), as is evident also from their published correspondence.

57. See also Fisher (1958, 1959, 1960). In a letter to D. J. Finney dated 15 March 1955, Fisher says he has "recently been thinking a little about the semantics" of the word "probability" (Bennett, 1990, p. 96).

58. Fisher's distaste for and suspicion of axiomatics is evident throughout his published correspondence; see, for example, Bennett (1990, pp. 128–129, 175, 185, and 331).

59. For Fisher's exchange with Jeffreys, see Lane (1980). (In addition to the papers discussed by Lane, there is also an exchange between Fisher and Jeffreys that occurs at the end of Fisher's 1935 *Journal of the Royal Statistical Society* paper.)

REFERENCES

Papers of Fisher referred to only on a single occasion are cited by their number in Fisher's *Collected Papers* (CP; Bennett, 1971–1974) and are not included below.

Barnard, G. (1963). Fisher's contributions to mathematical statistics. *J. Roy. Statist. Soc. Ser. A* **126** 162–166.

Barnard, G. (1990). Fisher: A retrospective. *Chance* **3** 22–28.

Bartlett, M. S. (1936). The information available in small samples. *Proceedings of the Cambridge Philosophical Society* **32** 560–566.

Bartlett, M. S. (1937). Properties of sufficiency and statistical tests. *Proc. Roy. Statist. Soc. Ser. A* **160** 268–282.

Bartlett, M. S. (1939). Complete simultaneous fiducial distributions. *Ann. Math. Statist.* **10** 129–138.

Bartlett, M. S. (1965). R. A. Fisher and the first fifty years of statistical methodology. *J. Amer. Statist. Assoc.* **60** 395–409.

Bennett, J. H., ed. (1971–1974). *Collected Papers of R. A. Fisher*. Adelaide, Aus.: Univ. Adelaide.

Bennett, J. H., ed. (1990). *Statistical Inference and Analysis: Selected Correspondence of R. A. Fisher*. Oxford, UK: Clarendon Press.

Box, J. F. (1978). *R. A. Fisher: The Life of a Scientist*. New York: Wiley.

Buehler, R. J. and Feddersen, A. P. (1963). Note on a conditional property of Student's t. *Ann. Math. Statist.* **34** 1098–1100.

CMS. See Fisher, 1950.

Creasey, M. A. (1954). Limits for the ratio of means. *J. Roy. Statist. Soc. Ser. B* **16** 186–194.

Dempster, A. P. (1964). On the difficulties inherent in Fisher's fiducial argument. *J. Amer. Statist. Assoc.* **59** 56–66.

Edwards, A. W. F. (1974). The history of likelihood. *Internat. Statist. Rev.* **42** 9–15.

Fienberg, S. E. (1980). Fisher's contributions to the analysis of categorical data. *R. A. Fisher: An Appreciation. Lecture Notes in Statist.* **1** 75–84. New York: Springer.

Fisher, R. A. (1915). Frequency distribution of the values of the correlation coefficient in samples from an indefinitely large population. *Biometrika* **10** 507–521. [CP 4.]

Fisher, R. A. (1918). On the correlation between relatives on the supposition of Mendelian inheritance. *Transactions of the Royal Society of Edinburgh* **52** 399–433. [CP 9.]

Fisher, R. A. (1921). On the "probable error" of a coefficient of correlation deduced from a small sample. *Metron* **1** 3–32. [CP 14.]

Fisher, R. A. (1922). On the mathematical foundations of theoretical statistics. *Philos. Trans. Roy. Soc. London Ser. A* **222** 309–368. [CP 18.]

Fisher, R. A. (1925). *Statistical Methods for Research Workers*. Edinburgh: Oliver and Boyd. [Many later editions.]

Fisher, R. A. (1930). Inverse probability. *Proceedings of the Cambridge Philosophical Society* **26** 528–535. [CP 84.]

Fisher, R. A. (1933). The concepts of inverse probability and fiducial probability referring to unknown parameters. *Proc. Roy. Soc. London Ser. A* **139** 343–348. [CP 102.]

Fisher, R. A. (1934). Two new properties of mathematical likelihood. *Proc. Roy. Soc. London Ser. A* **144** 285–307. [CP 108.]

Fisher, R. A. (1935a). The fiducial argument in statistical inference. *Annals of Eugenics* **6** 391–398. [CP 125.]

Fisher, R. A. (1935b). The logic of inductive inference (with discussion). *J. Roy. Statist. Soc.* **98** 39–82.

Fisher, R. A. (1935c). *The Design of Experiments*. Edinburgh: Oliver and Boyd. [Many later editions.]

Fisher, R. A. (1936). Uncertain inference. *Proceedings of the American Academy of Arts and Science* **71** 245–258. [CP 137.]

Fisher, R. A. (1937). On a point raised by M. S. Bartlett on fiducial probability. *Annals of Eugenics* **7** 370–375. [CP 151.]

Fisher, R. A. (1939a). The comparison of samples with possibly unequal variance. *Annals of Eugenics* **9** 174–180. [CP 162.]

Fisher, R. A. (1939b). A note on fiducial inference. *Ann. Math. Statist.* **10** 383–388. [CP 164.]

Fisher, R. A. (1939c). "Student." *Annals of Eugenics* **9** 1–9. [CP 165.]

Fisher, R. A. (1941). The asymptotic approach to Behrens's integral, with further tables for the *d* test of significance. *Annals of Eugenics* **11** 141–172. [CP 181.]

Fisher, R. A. (1945). The logical inversion of the notion of the random variable. *Sankhyā* **7** 129–132. [CP 203.]

Fisher, R. A. (1946). Testing the difference between two means of observations of unequal precision. *Nature* **158** 713. [CP 207.]

Fisher, R. A. (1948). Conclusions fiduciaires. *Ann. Inst. H. Poincaré* **10** 191–213. [CP 222.]

Fisher, R. A. (1950). *Contributions to Mathematical Statistics* [CMS]. Wiley, New York.

Fisher, R. A. (1951). Statistics. In *Scientific Thought in the Twentieth Century* (A. E. Heath, ed.) 31–55. London: Watts. [CP 242.]

Fisher, R. A. (1954). Contribution to a discussion of a paper on interval estimation by M. A. Creasy. *J. Roy. Statist. Soc. Ser. B* **16** 212–213.

Fisher, R. A. (1955). Statistical methods and scientific induction. *J. Roy. Statist. Soc. Ser. B* **17** 69–78. [CP 261.]

Fisher, R. A. (1956). *Statistical Methods and Scientific Inference* [SMSI]. New York: Hafner Press. [2nd ed., 1959; 3rd ed., 1973; Page references in the text are to the 3rd ed.]

Fisher, R. A. (1958). The nature of probability. *Centennial Review* **2** 261–274. [CP 272.]

Fisher, R. A. (1959). Mathematical probability in the natural sciences. *Technometrics* **1** 21–29. [CP 273.]

Fisher, R. A. (1960). Scientific thought and the refinement of human reasoning. *J. Oper. Res. Soc. Japan* **3** 1–10. [CP 282.]

Good, I. J. (1971). Reply to Professor Barnard. In *Foundations of Statistical Inference* (V. P. Godambe and D. A. Sprott, eds.) 138–140. Toronto: Holt, Rinehart, and Winston.

Hacking, I. (1990). *The Taming of Chance*. New York: Cambridge University Press.

Jeffreys, H. (1932). On the theory of errors and least squares. *Proc. Roy Soc. London Ser. A* **138** 38–45.

Jeffreys, H. (1939). *Theory of Probability*. Oxford, UK: Clarendon Press. [2nd ed., 1948; 3rd ed., 1961.]

Jeffreys, H. (1940). Note on the Behrens-Fisher formula. *Annals of Eugenics.* **6** 391–398.

Kendall, M. G. (1963). Ronald Aylmer Fisher, 1890–1962. *Biometrika* **50** 1–15.

Lane, D. (1980). Fisher, Jeffreys, and the nature of probability. *R. A. Fisher: An Appreciation. Lecture Notes in Statist.* **1** 148–160. New York: Springer.

Neyman, J. (1934). On the two different aspects of the representative method: The method of stratified sampling and the method of purposive selection. *J. Roy. Statist. Soc. Ser. A* **97** 558–625.

Neyman, J. (1935a). Statistical problems in agricultural experimentation (with K. Iwaszkiewicz and St. Kolodziejczyk). *J. Roy. Statist. Soc. B Suppl.* **2** 107–180.

Neyman, J. (1935b). On the problem of confidence intervals. *Ann. Math. Statist.* **6** 111–116.

Neyman, J. (1941). Fiducial argument and the theory of confidence intervals. *Biometrika* **32** 128–150.

Neyman, J. (1961). Silver jubilee of my dispute with Fisher. *J. Oper. Res. Soc. Japan* **3** 145–154.

Neyman, J. and Pearson, E. S. (1933). On the problem of the most efficient tests of statistical hypotheses. *Phil. Trans. Roy Soc. Ser. A* **231** 289–337.

Pearson, E. S. (1968). Some early correspondence between W. S. Gosset, R. A. Fisher and Karl Pearson, with notes and comments. *Biometrika* **55** 445–457.

Pearson, E. S. (1990). *'Student': A Statistical Biography of William Sealy Gosset* (R. L. Plackett and G. A. Barnard, eds.). Oxford, UK: Clarendon Press.

Pearson, K. (1892). *The Grammar of Science*. Walter Scott, London. [2nd ed., 1900; 3rd ed., 1911.]

Pearson, K. (1920). The fundamental problem of practical statistics. *Biometrika* **13** 1–16.

Pedersen, J. G. (1978). Fiducial inference. *Internat. Statist. Rev.* **46** 147–170.

Reid, C. (1982). *Neyman – From Life*. New York: Springer.

Savage, L. J. (1964). Discussion. *Bull. Inst. Internat. Statist.* **40** 925–927.

Savage, L. J. (1976). On re-reading R. A. Fisher (with discussion). *Ann. Statist.* **4** 441–500.

SMSI. See Fisher, 1956.

Soper, H. E., Young, A. W., Cave, B. H., Lee, A. and Pearson, K. (1917). A cooperative study. On the distribution of the correlation coefficient in small samples. Appendix II to the Papers of 'Student' and R. A. Fisher. *Biometrika* **11** 328–413.

Stone, M. (1983). Fiducial probability. *Encyclopedia of Statistical Sciences* **3** 81–86. New York: Wiley.

Wallace, D. (1980). The Behrens-Fisher and Fieller-Creasey problems. *R. A. Fisher: An Appreciation. Lecture Notes in Statist.* **1** 119–147. New York: Springer.

Wilkinson, G. N. (1977). On resolving the controversy in statistical inference (with discussion). *J. Roy. Statist. Soc. Ser. B* **39** 119–171.

Wilks, S. S. (1938). Fiducial distributions in fiducial inference. *Ann. Math. Statist.* **9** 272–280.

Yates, F. (1939). An apparent inconsistency arising from tests of significance based on fiducial distributions of unknown parameters. *Proceedings of the Cambridge Philosophical Society* **35** 579–591.

9

Alan Turing and the Central
Limit Theorem

Although the English mathematician Alan Mathison Turing (1912–1954) is remembered today primarily for his work in mathematical logic (Turing machines and the "Entscheidungsproblem"), machine computation, and artificial intelligence (the "Turing test"), his name is not usually thought of in connection with either probability or statistics. One of the basic tools in both of these subjects is the use of the normal or Gaussian distribution as an approximation, one basic result being the Lindeberg-Feller central limit theorem taught in first-year graduate courses in mathematical probability. No one associates Turing with the central limit theorem, but in 1934 Turing, while still an undergraduate, rediscovered a version of Lindeberg's 1922 theorem and much of the Feller-Lévy converse to it (then unpublished). This paper discusses Turing's connection with the central limit theorem and its surprising aftermath: his use of statistical methods during World War II to break key German military codes.

1. INTRODUCTION

Turing went up to Cambridge as an undergraduate in the Fall Term of 1931, having gained a scholarship to King's College. (Ironically, King's was his second choice; he had failed to gain a scholarship to Trinity.) Two years later, during the course of his studies, Turing attended a series of lectures on the Methodology of Science, given in the autumn of 1933 by the distinguished astrophysicist Sir Arthur Stanley Eddington. One topic Eddington discussed was the tendency of experimental measurements subject to errors of observation to often have an approximately normal or Gaussian distribution. But Eddington's heuristic sketch left Turing dissatisfied; and Turing set out to derive a rigorous mathematical proof of what is today termed the central limit

Reprinted with permission from *American Mathematical Monthly* 102, no. 6 (1995): 483–494.

theorem for independent (but not necessarily identically distributed) random variables.

Turing succeeded in his objective within the short span of several months (no later than the end of February 1934). Only then did he find out that the problem had already been solved, twelve years earlier, in 1922, by the Finnish mathematician Jarl Waldemar Lindeberg (1876–1932). Despite this, Turing was encouraged to submit his work, suitably amended, as a Fellowship Dissertation. (Turing was still an undergraduate at the time; students seeking to become a Fellow at a Cambridge college had to submit evidence of original work, but did not need to have a Ph.D. or its equivalent.) This revision, entitled "On the Gaussian Error Function," was completed and submitted in November, 1934. On the strength of this paper Turing was elected a Fellow of King's four months later (March 16, 1935) at the age of 22; his nomination supported by the group theorist Philip Hall and the economists John Maynard Keynes and Alfred Cecil Pigou. Later that year the paper was awarded the prestigious Smith's prize by the University (see Hodges, 1983).

Turing never published his paper. Its major result had been anticipated, although, as will be seen, it contains other results that were both interesting and novel at the time. But in the interim Turing's mathematical interests had taken a very different turn. During the spring of 1935, awaiting the outcome of his application for a Fellowship at King's, Turing attended a course of lectures by the topologist M. H. A. Newman on the Foundations of Mathematics. During the International Congress of Mathematicians in 1928, David Hilbert had posed three questions: is mathematics *complete* (that is, can every statement in the language of number theory be either proved or disproved?), is it *consistent*, and is it *decidable*? (This last is the *Entscheidungsproblem*, or the "decision problem"; does there exist an *algorithm* for deciding whether or not a specific mathematical assertion does or does not have a proof.) Kurt Gödel had shown in 1931 that the answer to the first question is *no* (the so-called "first incompleteness theorem"); and that if number theory is consistent, then a proof of this fact does not exist using the methods of the first-order predicate calculus (the "second incompleteness theorem"). Newman had proved the Gödel theorems in his course, but he pointed out that the third of Hilbert's questions, the *Entscheidungsproblem*, remained open.

This challenge attracted Turing, and in short order he had arrived at a solution (in the negative), using the novel device of *Turing machines*. The drafting of the resulting paper (Turing, 1937), dominated Turing's life for a year from the Spring of 1935 (Hodges, 1983, p. 109); and thus Turing turned from mathematical probability, never to return.

A copy of Turing's Fellowship Dissertation survives, however, in the archives of the King's College Library; and its existence raises an obvious question. Just how far did a mathematician of the calibre of Turing get in this attack on the central limit theorem, one year before he began his pioneering research into the foundations of mathematical logic? The answer to that question is the focus of this paper.

2. THE CENTRAL LIMIT THEOREM

The earliest version of the central limit theorem (CLT) is due to Abraham de Moivre (1667–1754). If $X_1, X_2, X_3, \ldots$ is an infinite sequence of 1's and 0's recording whether a success $(X_n = 1)$ or failure $(X_n = 0)$ has occurred at each stage in a sequence of repeated trials, then the sum $S_n =: X_1 + X_2 + \cdots + X_n$ gives the total number of successes after n trials. If the trials are independent, and the probability of a success at each trial is the same, say $P[X_n = 1] = p$, $P[X_n = 0] = 1 - p$, then the probability of seeing exactly k successes in n trials has a binomial distribution:

$$P[S_n = k] = \frac{n!}{k!(n-k)!} p^k (1-p)^{n-k}.$$

If n is large (for example, 10,000), then as de Moivre noted, the direct computation of binomial probabilities "is not possible without labor nearly immense, not to say impossible"; and for this reason he turned to approximate methods (see Diaconis and Zabell, 1991): using Stirling's approximation (including correction terms) to estimate the individual terms in the binomial distribution and then summing, de Moivre discovered the remarkable fact that

$$\lim_{n \to \infty} P\left[a \le \frac{S_n - np}{\sqrt{np(1-p)}} \le b \right] = \frac{1}{\sqrt{2\pi}} \int_a^b \exp\left[-\frac{1}{2}x^2 \right] dx,$$

or $\Phi(b) - \Phi(a)$, where $\Phi(x)$ is the cumulative distribution function of the standard normal (or Gaussian) distribution:

$$\Phi(x) =: \frac{1}{\sqrt{2\pi}} \int_{-\infty}^x \exp\left[-\frac{1}{2}t^2 \right] dt.$$

During the 19th and 20th centuries this result was extended far beyond the simple coin-tossing setup considered by de Moivre, important contributions being made by Laplace, Poisson, Chebyshev, Markov, Liapunov, von Mises, Lindeberg, Lévy, Bernstein, and Feller; see Adams (1974), Maistrov (1974), Le Cam (1986), and Stigler (1986) for further historical information. Such investigations revealed that if $X_1, X_2, X_3, \ldots$ is *any* sequence of independent random variables having the same distribution, then the sum S_n satisfies the

201

CLT provided suitable centering and scaling constants are used: the centering constant np in the binomial case is replaced by the sum of the *expectations* $E[X_i]$; the scaling constant $\sqrt{np(1-p)}$ is replaced by the square root of the sum of the *variances* $\mathrm{Var}[X_i]$ (provided these are finite).

Indeed, it is not even necessary for the random variables X_n contributing to the sum S_n to have the same distribution, provided that no one term dominates the sum. Of course this has to be made precise. The best result is due to Lindeberg. Suppose $E[X_n] = 0$, $0 < \mathrm{Var}[X_n] < \infty$, $s_n^2 =: \mathrm{Var}[S_n]$, and

$$\Lambda_n(\varepsilon) =: \sum_{k=1}^{n} E\left[\left(\frac{X_k}{s_n} \right)^2 ; \frac{|X_k|}{s_n} \geq \varepsilon \right].$$

(The notation $E[X; Y \geq \varepsilon]$ means the expectation of X is restricted to outcomes ω such that $Y(\omega) \geq \varepsilon$.) The *Lindeberg condition* is the requirement that

$$\Lambda_n(\varepsilon) \to 0, \quad \forall \varepsilon > 0; \tag{2.1}$$

and the *Lindeberg central limit theorem* (Lindeberg, 1922) states that if the sequence of random variables $X_1, X_2, \ldots$ satisfies the Lindeberg condition (2.1), then for all $a < b$,

$$\lim_{n \to \infty} P\left[a < \frac{S_n}{s_n} < b \right] = \Phi(b) - \Phi(a). \tag{2.2}$$

Despite its technical appearance, the Lindeberg condition turns out to be a natural sufficient condition for the CLT. There are two reasons for this. First, the Lindeberg condition has a simple consequence: if $\sigma_k^2 =: \mathrm{Var}[X_k]$, then

$$\rho_n^2 =: \max_{k \leq n} \left(\frac{\sigma_k^2}{s_n^2} \right) \to 0. \tag{2.3}$$

Thus, if the sequence $X_1, X_2, X_3, \ldots$ satisfies the Lindeberg condition, the variance of an individual term X_k in the sum S_n is asymptotically negligible. Second, for such sequences the Lindeberg condition is *necessary* as well as sufficient for the CLT to hold, a beautiful fact discovered (independently) by William Feller and Paul Lévy in 1935. In short: $(2.1) \leftrightarrow (2.2) + (2.3)$.

If, in contrast, the Feller-Lévy condition (2.3) fails, then it turns out that convergence to the normal distribution can occur in a fashion markedly different from that of the CLT. If (2.3) does *not* hold, then there exists a number $\rho > 0$, and two sequences of positive integers $\{m_k\}$ and $\{n_k\}$, $\{n_k\}$ strictly increasing, such that

$$1 \leq m_k \leq n_k \quad \text{for all } k \quad \text{and} \quad \mathrm{Var}\left[\frac{X_{m_k}}{s_{n_k}} \right] = \frac{\sigma_{m_k}^2}{s_{n_k}^2} \to \rho^2 > 0. \tag{2.4}$$

202

Feller (1937) showed that if normal convergence occurs (that is, condition (2.2) holds), but condition (2.4) also obtains, then

$$\frac{1}{\rho}\frac{X_{m_k}}{s_{n_k}} \Rightarrow N(0, 1).$$

That is, there exists a subsequence X_{m_k} whose contributions to the sums S_n are nonnegligible (relative to s_n) and which, properly scaled, converges to the standard normal distribution. (The symbol "$\Rightarrow$" denotes convergence in distribution; $N(\mu, \sigma^2)$ the normal distribution having expectation μ, variance σ^2.)

Note. For the purposes of brevity, this summary of the contributions of Feller and Lévy simplifies a much more complex story; see Le Cam (1986) for a more detailed account. (Or better, consult the original papers themselves!)

3. TURING'S FELLOWSHIP DISSERTATION

Turing's fellowship dissertation was written twelve years after Lindeberg's work had appeared, and shortly before the work of Feller and Lévy. There are several aspects of the paper that demonstrate Turing's insight into the basic problems surrounding the CLT. One of these is his decision, contrary to a then common textbook approach (e.g., Burnside, 1928, pp. 87–90), but crucial if the best result is to be obtained (and the approach also adopted by Lindeberg), to work at the level of distribution functions (i.e., the function $F_X(t) =: P[X \leq t]$) rather than densities (the derivatives of the distribution functions). In Appendix B Turing notes:

I have attempted to obtain some results [using densities] . . . but without success. The reason is clear. In order that the shape frequency functions $u_n(x)$ of $f_n(x)$ should tend to the shape frequency function $\phi(x)$ of the Gaussian error much heavier restrictions on the functions $g_n(x)$ are required than is needed if we only require that $U_n \rightarrow \Phi$. It became clear to me . . . that it would be better to work in terms of distribution function throughout.

This was an important insight. Although versions of the central limit theorem do exist for densities, these ordinarily require stronger assumptions than just the Lindeberg condition (2.1); see, e.g., Feller (1971), pp. 516–517, Petrov (1975), Chapter 7.

Let us now turn to the body of Turing's paper, and consider it, section by section.

3.1. Basic Structure of the Paper

The first seven sections of the paper (pp. 1–6) summarize notation and the basic properties of distribution functions. Section 1 summarizes the problem; Section 2 defines the distribution function F (abbreviated DF) of an "error" ε; Section 3 summarizes the basic properties of the expectation and mean square deviation (MSD) of a sum of independent errors; rigorous proofs in terms of the distribution function are given in an appendix at the end of the paper (Appendix C). Section 4 discusses the distribution function of a sum of independent errors, the *sum distribution function* (SDF), in terms of the distribution functions of each term in the sum, and derives the formula for $F \oplus G$, the convolution of two distribution functions. Section 5 then introduces the concept of the *shape function* (SF); the standardization of a distribution function F to have zero expectation and unit MSD; thus, if F has expectation μ and MSD $\sigma^2 (\sigma > 0)$, then the shape function of F is $U(x) =: F(\sigma(x - \mu))$. (Turing uses the symbols "a" and "k^2" to denote μ and σ^2; several other minor changes in notation of this sort are made below.)

In Section 6 Turing then states the basic problem to be considered: given a sequence of errors ε_k, having distribution functions G_k, shape functions V_k, means μ_k, mean square deviations σ_k^2, sum distribution functions F_n, and shape functions U_n for each F_n, under what conditions do the shape functions $U_n(x)$ converge uniformly to $\Phi(x)$, the "SF of the Gaussian Error"? Turing then assumes for simplicity that $\mu_k = 0$ and $\sigma_k^2 < \infty$. In Section 7 ("Fundamental Property of the Gaussian Error"), he notes the only properties of Φ that are used in deriving sufficient conditions for normal convergence are that it is an SF, and the "self-reproductive property" of Φ: that is, if $X_1 \sim N(0, \sigma_1^2)$ and $X_2 \sim N(0, \sigma_2^2)$ are independent, then $X_1 + X_2 \sim N(0, \sigma_1^2 + \sigma_2^2)$. (The notation "$X \sim N(\mu, \sigma^2)$" means that the random variable X has the distribution $N(\mu, \sigma^2)$.)

3.2. The Quasi-Necessary Conditions

It is at this point that Turing comes to the heart of the matter. In Section 8 ("The Quasi-Necessary Conditions") Turing notes

> The conditions we shall impose fall into two groups. Those of one group (the quasi-necessary conditions) involve the MSDs only. They are not actually necessary, but if they are not fulfilled U_n can only tend to Φ by a kind of accident.

The two conditions that Turing refers to as the "quasi-necessary" conditions

204

are:

$$\sum_{k=1}^{\infty} \sigma_k^2 = \infty \quad \text{and} \quad \frac{\sigma_n^2}{s_n^2} \to 0. \tag{3.1}$$

It is easy to see that Turing's condition (3.1) is equivalent to condition (2.3). (That (2.3) $\Rightarrow$ (3.1) is immediate. To see (3.1) $\Rightarrow$ (2.3): given $\varepsilon > 0$, choose $M \geq 1$ so that $\sigma_n^2/s_n^2 < \varepsilon$ for $n \geq M$, and $N \geq M$ so that $\sigma_k^2/s_N^2 < \varepsilon$ for $1 \leq k \leq M$; if $n \geq N$, then $\sigma_k^2/s_n^2 < \varepsilon$ for $1 \leq k \leq n$.)

In his Theorems 4 and 5, Turing explores the consequences of the failure of either part of condition (3.1). Turing's proof of Theorem 4 requires his

Theorem 3. *If X and Y are independent, and both X and $X + Y$ are Gaussian, then Y is Gaussian.*

This is a special case of a celebrated theorem proven shortly thereafter by Harald Cramér (1936); if X and Y are independent, and $X + Y$ is Gaussian, then both X and Y must be Gaussian. Lévy had earlier conjectured Cramér's theorem to be true (in 1928 and again in 1935), but had been unable to prove it. Cramér's proof of this result in 1936 in turn enabled Lévy to arrive at necessary and sufficient conditions for the CLT of a very general type (using centering and scaling constants other than the mean and standard deviation), and this in turn led Lévy to write his famous monograph, *Théorie de l'Addition des Variables Aléatoires* (Lévy, 1937); see Le Cam (1986, pp. 80–81, 90).

Cramér's theorem is a hard fact; his original proof appealed to Hadamard's theorem in the theory of entire functions. The special case of the theorem needed by Turing is much simpler; it is an immediate consequence of the characterization theorem for characteristic functions. To see this, let $\phi_X(t) =:$ $E[\exp(itX)]$ denote the characteristic function of a random variable X; and suppose that X and Y are independent, $X \sim N(0, \sigma^2)$, and $X + Y \sim N(0, \sigma^2 + \tau^2)$. Then

$$\exp\left(-\frac{\sigma^2 + \tau^2}{2}t^2\right) = \phi_{X+Y}(t) = \phi_x(t)\phi_Y(t) = \exp\left(-\frac{\sigma^2}{2}t^2\right)\phi_Y(t),$$

hence $\phi_Y(t) = \exp(-(\tau^2/2)t^2)$; thus $Y \sim N(0, \tau^2)$ because the characteristic function of a random variable uniquely determines the distribution of that variable. Turing's proof, which uses distribution functions, is not much longer.

It is an immediate consequence of Cramér's theorem that if $S_n/s_n \Rightarrow$ $N(0, 1)$, but $\lim_{n \to \infty} s_n^2 < \infty$, then all the summands X_j must in fact be Gaussian. But Turing did not have this fact at his disposal, only his much weaker Theorem 3. His **Theorem 4** (phrased in the language of random variables) thus makes the much more limited claim that if (a) $\Sigma \sigma_n^2 < \infty$, (b) S_n converges to a Gaussian distribution, and (c) X_0 is a random variable at once independent of the original sequence $X_1, X_2, \ldots$ and having a distribution other than Gaussian, then the sequence $S_n^* = X_0 + S_n$ *cannot* converge to the Gaussian distribution. In other words: if $\Sigma \sigma_n^2 < \infty$, then "the convergence . . . to the Gaussian is so delicate that a single extra term in the sequence . . . upsets it" (p. 17).

Turing's **Theorem 5** in turn explores the consequences of the failure of (3.1) in the case that $\Sigma \sigma_n^2 = \infty$, but $\rho_n^2 =: \sigma_n^2 / s_n^2$ does not tend to zero as $n \to \infty$. The statement of the theorem is somewhat technical in nature, but Turing's later summary of it captures the essential phenomenon involved:

If F_n [the distribution function of S_n] tends to Gaussian and σ_n^2 / s_n^2 does not tend to zero [but $\Sigma \sigma_n^2 = \infty$] we can find a subsequence of G_n [the distribution function of X_n] tending to Gaussian.

Thus Turing had by some two years anticipated Feller's discovery of the subsequence phenomenon. (In Turing's typescript, symbols such as "F_n" are entered by hand; in the above quotation the space for "F_n" has by accident been left blank, but the paragraph immediately preceding this one in the typescript makes it clear that "F_n" is intended.)

3.3. The Sufficient Conditions

Turing states in his preface that he had been "informed that an almost identical proof had been given by Lindeberg." This comment refers to the *method* of proof Turing uses, not the *result* obtained. Turing's method is to smooth the distribution functions $F_n(x)$ of the sum by forming the convolution $F_n *$ $\Phi(x/\rho)$, expand the result in a Taylor series to third order, and then let the variance ρ^2 of the convolution term tend to zero. This is similar to the method employed by Lindeberg. (There is an important difference, however: Turing does not use Lindeberg's "swapping" argument. For an attractive modern presentation of the Lindeberg method, see Breiman, 1968, pp. 167–170; for discussion of the method, Pollard's comments in Le Cam, 1986, pp. 94–95.)

Turing does *not*, however, succeed in arriving at the Lindeberg condition (2.1) as a sufficient condition for convergence to the normal distribution; the most general sufficient condition he gives (on p. 27) is complex in appearance (although it necessarily implies the Lindeberg condition). Turing concedes that his "form of the sufficiency conditions is too clumsy for direct application," but notes that it can be used to "derive various criteria from it, of different degrees of directness and of comprehensiveness" (p. 28). One of

206

these holds if the summands X_k all have the same *shape* (that is, the shape functions $V_k(x) =: P[X_k/\sigma_k \leq x]$ coincide); and thus includes the special case of identically distributed summands having a second moment. (This was no small feat, since even this special case of the more general Lindeberg result had eluded proof until the publication of Lindeberg's paper.)

One formulation of this criterion, equivalent to the one actually stated by Turing, is: there exists a function $J: \mathbf{R}^+ \to \mathbf{R}^+$ such that $\lim_{t \to \infty} J(t) = 0$, and

$$E\left[\left(\frac{X_k}{\sigma_k} - t\right)^2 ; \left|\frac{X_k}{\sigma_k}\right| \geq t\right] \leq J(t) \quad \text{for all } k \geq 1, t \geq 0. \quad (3.2)$$

In turn one simple sufficient condition for this given by Turing (pp. 30–31) is that there exists a function ϕ such that $\phi(x) > 0$ for all x, $\lim_{x \to \pm\infty} \phi(x) = \infty$, and

$$\sup_k E\left[\left(\frac{X_k}{\sigma_k}\right)^2 \phi\left(\frac{X_k}{\sigma_k}\right)\right] < \infty. \quad (3.3)$$

(Note that unfortunately one important special case not covered by either of these conditions is that the X_k are *uniformly bounded*: $|X_k| \leq C$ for some C and all $k \geq 1$.)

In assessing this portion of Turing's paper, it is important to keep two points in mind. First, Turing states in his preface that "since reading Lindeberg's paper I have for obvious reasons made no alterations to that part of the paper which is similar to his." The manuscript is thus necessarily incomplete; it presumably would have been further polished and refined had Turing continued to work on it; the technical sufficient conditions given represent how far Turing had gotten on the problem *prior* to seeing Lindeberg's work. Second, in 1934 the Lindeberg condition was only known to be *sufficient*, not necessary; thus even in discussing his results in other sections of the paper (where he felt free to refer to the Lindeberg result), it may not have seemed important to Turing to contrast his own particular technical sufficient conditions with those of Lindeberg; the similarity in method must have seemed far more important.

3.4. One Counterexample

In Section 14 Turing concludes by giving a simple example of a sequence $X_1, X_2, \ldots$ that satisfies the quasi-necessary conditions (3.1), but not the CLT.

207

For $n \geq 1$, let
$$P[X_n = \pm n] = \frac{1}{2n^2}; \quad P[X_n = 0] = 1 - \frac{1}{n^2}.$$
Then $E[X_n] = 0$, $\mathrm{Var}[X_n] = E[X_n^2] = 1$, $s_n^2 = \mathrm{Var}[S_n] = n \to \infty$, and $\rho_n^2 = 1/n \to 0$; thus (3.1) is satisfied. Turing then shows that if S_n/s_n converges, the limit distribution must have a discontinuity at zero, and therefore cannot be Gaussian.

It is interesting that Turing should happen to choose this particular example; although he does not note it, the sequence $\{S_n/s_n: n \geq 1\}$ has the property that $\mathrm{Var}[S_n/s_n] \equiv 1$, but $\lim_{n \to \infty} S_n(\omega)/s_n = 0$ for almost all sample paths ω. This is an easy consequence of the first Borel-Cantelli lemma: because
$$\sum_{n=1}^{\infty} P[X_n \neq 0] = \sum_{n=1}^{\infty} \frac{1}{n^2} = \zeta(2) = \frac{\pi^2}{6} < \infty,$$
it follows that $P[X_n \neq 0 \text{ infinitely often}] = 0$; thus $P[\sup_n |S_n| < \infty] = 1$ and $P[\lim_{n \to \infty} S_n/s_n = 0] = 1$.

The existence of such sequences has an interesting consequence for the CLT. Let $\{Y_n: n \geq 1\}$ be a sequence of independent random variables, jointly independent of the sequence $\{X_n: n \geq 1\}$ and such that $P[Y_n = \pm 1] = \frac{1}{2}$. Let $T_n =: Y_1 + Y_2 + \cdots + Y_n$; then a trite calculation shows that $S_n + T_n$ satisfies the Feller condition (2.3), but not the Lindeberg condition (2.1). Let $t_n^2 =: \mathrm{Var}[T_n]$; then $T_n/t_n \Rightarrow N(0, 1)$ and $\mathrm{Var}[S_n + T_n] = s_n^2 + t_n^2$, hence

$$\frac{S_n + T_n}{\sqrt{\mathrm{Var}[S_n + T_n]}} = \frac{s_n}{\sqrt{s_n^2 + t_n^2}} \left(\frac{S_n}{s_n} \right) + \frac{t_n}{\sqrt{s_n^2 + t_n^2}} \left(\frac{T_n}{t_n} \right)$$

$$= \left(\frac{1}{\sqrt{2}} \right) \left(\frac{S_n}{s_n} \right) + \left(\frac{1}{\sqrt{2}} \right) \left(\frac{T_n}{t_n} \right)$$

$$\Rightarrow N \left(0, \tfrac{1}{2} \right).$$

Thus the sequence $S_n + T_n$ does converge to a Gaussian distribution! This does not, however, contradict the Feller converse to the Lindeberg CLT; that result states that $S_n + T_n$, rescaled to have unit variance, cannot converge to the *standard* Gaussian $N(0, 1)$.

4. DISCUSSION

Turing's Fellowship Dissertation tells us something about Turing, something about the state of mathematical probability at Cambridge in the 1930s, and something about the general state of mathematical probability during that decade.

I. J. Good (1980, p. 34) has remarked that when Turing "attacked a problem he started from first principles, and he was hardly influenced by received opinion. This attitude gave depth and originality to his thinking, and also it helped him to choose important problems." This observation is nicely illustrated by Turing's work on the CLT. His dissertation is, viewed in context, a very impressive piece of work. Coming to the subject as an undergraduate, his knowledge of mathematical probability was apparently limited to some of the older textbooks such as "Czuber, Morgan Crofton, and others" (*Preface*, p. ii). Despite this, Turing immediately realized the importance of working at the level of distribution functions rather than densities; developed a method of attack similar to Lindeberg's; obtained useful sufficient conditions for convergence to the normal distribution; identified the conditions necessary for true central limit behavior to occur; understood the relevance of a Cramér-type factorization theorem in the derivation of such necessary conditions; and discovered the Feller subsequence phenomenon. If one realizes that the defects of the paper, such as they are, must largely reflect the fact that Turing had ceased to work on the main body of it after being apprised of Lindeberg's work, it is clear that Turing had penetrated almost immediately to the heart of a problem whose solution had long eluded many mathematicians far better versed in the subject than he. (It is interesting to note that Lindeberg was also a relative outsider to probability theory, and only began to work in the field a few years before 1922.)

The episode also illustrates the surprisingly backward state of mathematical probability in Cambridge at the time. Turing wrote to his mother in April, 1934: "I am sending some research I did last year to Czuber in Vienna [the author of several excellent German textbooks on mathematical probability], not having found anyone in Cambridge who is interested in it. I am afraid however that he may be dead, as he was writing books in 1891" (Hodges, 1983, p. 88). (Czuber had in fact died nearly a decade before, in 1925.)

This disinterest is particularly surprising in the case of G. H. Hardy, who was responsible for a number of important results in probabilistic number theory. But anyone who has studied the Hardy-Ramanujan proof of the distribution of prime divisors of an integer (1917), and compared it to Turán's (see Kac, 1959, pp. 71–74) will realize at once that even the most rudimentary ideas of modern probability must have been foreign to Hardy; see also Elliott (1979, pp. 1–5), Elliott (1980, pp. 16–20). Indeed, Paul Erdős believes that "had Hardy known the even least little bit of probability, with his amazing talent he would certainly have been able to prove the law of the iterated logarithm" (Diaconis, 1993). Perhaps this reflected in part the limited English literature on the subject. In 1927, when Harald Cramér visited England and mentioned to Hardy (his friend and former teacher) that he had become interested in probability theory, Hardy replied that "there was no mathematically satisfactory book in English on this subject, and encouraged me to write one" (Cramér, 1976, p. 516).

Finally, Turing's thesis illustrates the transitional nature of work in mathematical probability during the decade of the 1930s, before the impact of Kolmogorov's pioneering book *Grundbegriffe der Wahrscheinlichkeitsrechnung* (Kolmogorov, 1933) had been felt. In his paper Turing had thought it necessary to state and prove some of the most basic properties of distribution functions and their convolutions (in Sections 3 and 4, and Appendix C of the dissertation). His comment that his Appendix C "is only given for the sake of logical completeness and it is of little consequence whether it is original or not" (*Preface*, p. i), illustrates that such results, although "known," did not enjoy general currency at the time. (It is all too easy to overlook today the important milestone in the literature of the subject marked by the publication in 1946 of Harald Cramér's important textbook *Mathematical Methods of Statistics*.)

It is also interesting to note Turing's approach to the problem in terms of convolutions of distribution functions rather than sums of independent random variables. Feller had similarly avoided the use of the language of random variables in his 1935 paper, formulating the problem instead in terms of convolutions. The reason, as Le Cam (1986, p. 87) notes, was that "Feller did not think that such concepts [as random variable] belonged in a mathematical framework. This was a common attitude in the mathematical community."

Current mathematical attitudes toward probability have changed so markedly from the distrust and scepticism of earlier times that today the sheer magnitude of the shift is often unappreciated. Joseph Doob, whose own work dates back to this period, notes that "even as late as the 1930s it was not quite obvious to some probabilists, and it was certainly a matter of doubt to most nonprobabilists, that probability could be treated as a rigorous mathematical discipline. In fact it is clear from their publications that many probabilists were uneasy in their research until their problems were rephrased in what was then nonprobabilistic language" (Le Cam, 1986, pp. 93–94).

5. EPILOGUE: BLETCHLEY PARK

After his fellowship dissertation Turing "always looked out for any statistical aspects of [a] problem under consideration" (Britton, 1992, p. ix). This trait of Turing is particularly striking in the case of his cryptanalytic work during the Second World War.

Turing left England for Princeton in 1936, to work with the logician Alonzo Church; he returned in 1938, after his Fellowship at King's College had been renewed. Recruited almost immediately by GC and CS (the Government Code and Cipher School), on September 4th, 1939 (one day after the outbreak of war) Turing reported to Bletchley Park, the British cryptanalytic unit charged with breaking German codes, soon rising to a position of considerable importance. (Turing's work at Bletchley was the subject of a 1987 London play, "Breaking the Code," written by Hugh Whitemore and starring Derek Jacobi, of "I, Claudius" fame.)

The staff at Bletchley Park included many gifted people, distinguished in a number of different fields; among these were the mathematicians M. H. A. Newman, J. H. C. Whitehead, Philip Hall, Peter Hilton, Shaun Wylie, David Rees, and Gordon Welchman; the international chessmasters C. H. O'D. Alexander, P. S. Milner-Barry, and Harry Golombek; and others such as Donald Mitchie (today an important figure in artificial intelligence), Roy Jenkins (the later Chancellor of the Exchequer), and Peter Benenson (the founder of Amnesty International). Turing's chief statistical assistant in the later half of 1942 was another mathematician, I. J. Good, fresh from studies under Hardy and Besicovitch at Cambridge. (Good arrived at Bletchley on May 27, 1941, the day the *Bismarck* was sunk.) In recent years Good has written several papers (Good 1979, 1980, 1992, 1993a) discussing Turing's *ad hoc* development of Bayesian statistical methods at Bletchley to assist in the decrypting of German messages. (More general accounts of the work at Bletchley include Lewin, 1978, Welchman, 1982, and Hinsley and Stripp, 1993; see also the bibliography in Good, 1992.)

The specific details of Turing's statistical contributions are too complex to go into here. (Indeed, much of this information was until recently still classified and, perhaps for this reason, Good's initial papers on the subject do not even describe the specific cryptanalytic techniques developed by Turing; they give instead only a general idea of the type of statistical methods used. But in his most recent paper on this subject (Good, 1993a), Jack Good does provide a detailed picture of the various cryptanalytic techniques that Turing developed at Bletchley Park.) Three of Turing's most important statistical contributions were: (1) his discovery, independently of Wald, of some form of sequential analysis; (2) his anticipation of empirical Bayes methods (later further developed in the 1950s by Good and independently by Herbert Robbins); and (3) his use of logarithms of the Bayes factor (termed by Good the "weight of evidence") in the evaluation and execution of decryption. (For many references to the concept of weight of evidence, see, for example, Good, 1993b and the two indices of

Good, 1983.) The units for the logarithms, base 10, were termed *bans* and *decibans*:

The reason for the name ban was that tens of thousands of sheets of paper were printed in the town of Banbury on which weights of evidence were entered in decibans for carrying out an important process called Banburismus.... [Good, 1979, p. 394]

"Tens of thousands of sheets of paper...." This sentence makes it clear that Turing's contributions in this area were not mere idle academic speculation, but an integral part of the process of decryption employed at Bletchley.

One episode is particularly revealing as to the importance with which the Prime Minister, Winston Churchill, viewed the cryptanalytic work at Bletchley. On October 21, 1941, frustrated by bureaucratic inertia, Turing, Welchman, Alexander, and Milner-Barry wrote a letter *directly* to Churchill (headed "Secret and Confidential; Prime Minister only") complaining that inadequate personnel had been assigned to them; immediately upon its receipt Churchill sent a memo to his principal staff officer directing him to "make sure they have all they want on extreme priority and report to me that this had been done" (Hodges, 1983, pp. 219–221).

Much of I. J. Good's own work in statistics during the decades immediately after the end of the war was a natural outgrowth of his cryptanalytic work during it; this includes both his 1950 book *Probability and the Weighing of Evidence*; and his papers on the sampling of species (e.g., Good, 1953) and the estimation of probabilities in large sparse contingency tables (much of it summarized in Good, 1965). Some of this work was stimulated either directly (e.g., Good, 1973, p. 936) or indirectly (the influence being somewhat remote, however, in the case of contingency tables) by Turing's ideas:

Turing did not publish these war-time statistical ideas because, after the war, he was too busy working on the ground floor of computer science and artificial intelligence. I was impressed by the importance of his statistical ideas, for other applications, and developed and published some of them in various places. [Good, 1992, p. 211]

ACKNOWLEDGMENTS

I thank Anthony Edwards for his assistance in obtaining a copy of the typescript of Turing's Fellowship Dissertation during a visit to Cambridge in May 1992; and the Master and Fellows of Gonville and Caius College for their hospitality during that visit. Quotations from Turing's unpublished Fellowship Dissertation appear here by the kind permission of Professor Robin Gandy of Oxford University. Thanks also to Persi Diaconis, John Ewing, Jack Good,

Steve Stigler, and an anonymous referee for helpful comments on an earlier draft of the paper.

REFERENCES

Adams, W. J. (1974). *The Life and Times of the Central Limit Theorem*. New York: Kaedmon.

Breiman, L. (1968). *Probability*. Reading, MA: Addison-Wesley.

Britton, J. L., ed. (1992). *The Collected Works of A. M. Turing: Pure Mathematics*. Amsterdam: North-Holland. [Contains the two-page Preface to Turing's Fellowship Dissertation.]

Burnside, W. (1928). *Theory of Probability*. Cambridge University Press.

Cramér, H. (1936). Ueber eine Eigenschaft der normalen Verteilungsfunktion. *Mathematische Zeitschrift* 41, 405–414.

Cramér, H. (1946). *Mathematical Methods of Statistics*. Princeton, NJ: Princeton University Press.

Cramér, H. (1976). Half of a century of probability theory: some personal recollections. *Annals of Probability* 4, 509–546.

Diaconis, P. (1993). Personal communication. [The quotation is a paraphrase from memory.]

Diaconis, P. and Zabell, S. (1991). Closed form summation for classical distributions: variations on a theme of De Moivre. *Statistical Science* 6, 284–302.

Elliott, P. D. T. A. (1979). *Probabilistic Number Theory I: Mean Value Theorems*. New York: Springer-Verlag.

Elliott, P. D. T. A. (1980). *Probabilistic Number Theory II: Central Limit Theorems*. New York: Springer-Verlag.

Feller, W. (1935). Über den zentralen Grenzwertsatz der Wahrscheinlichkeitsrechnung. *Mathematische Zeitschrift* 40, 521–559.

Feller, W. (1937). Über den zentralen Grenzwertsatz der Wahrscheinlichkeitsrechnung, II. *Mathematische Zeitschrift* 42, 301–312.

Feller, W. (1971). *An Introduction to Probability Theory and Its Applications*, vol. 2, 2nd ed. New York: Wiley.

Good, I. J. (1953). The population frequencies of species and the estimation of population parameters. *Biometrika* 40, 237–264.

Good, I. J. (1965). *The Estimation of Probabilities: An Essay on Modern Bayesian Methods*. Cambridge, MA: M.I.T. Press.

Good, I. J. (1973). The joint probability generating function for run-lengths in regenerative binary Markov chains, with applications. *Annals of Statistics* 1, 933–939.

Good, I. J. (1979). A. M. Turing's statistical work in World War II, *Biometrika*, 66, 393–396.

Good, I. J. (1980). Pioneering work on computers at Bletchley. *A History of Computing in the Twentieth Century*, N. Metropolis, J. Howlett, and G.-C. Rota, (eds.) New York: Academic Press, pp. 31–45.

Good, I. J. (1983). *Good Thinking*. Minneapolis, MN: Minnesota University Press.

Good, I. J. (1992). Introductory remarks for the article in *Biometrika* 66 (1979). In *The Collected Works of A. M. Turing: Pure Mathematics* (J. L. Britton, ed.), Amsterdam: North-Holland, pp. 211–223.

213

Good, I. J. (1993a). Enigma and Fish. In *Codebreakers: The Inside Story of Bletchley Park* (F. H. Hinsley and A. Stripp, eds.), Oxford: Oxford University Press, pp. 149–166.

Good, I. J. (1993b). Causal tendency, necessitivity and sufficientivity: an updated review. In *Patrick Suppes, Scientific Philosopher* (P. Humphreys, ed.), Dordrecht: Kluwer.

Hardy, G. H. and Ramanujan, S. (1917). The normal number of prime factors of a number. *Quarterly J. Math.* 48, 76–92.

Hodges, A. (1983). *Alan Turing: The Enigma.* New York: Simon and Schuster.

Kac, M. (1959). *Statistical Independence in Probability, Analysis and Number Theory.* Carus Mathematical Monographs, Number 12. Mathematical Association of America.

Kolmogorov, A. A. (1933). *Grundbegriffe der Wahrscheinlichkeitsrechnung.* Ergebnisse der Mathematik, Berlin: Springer-Verlag.

Le Cam, L. (1986). The central limit theorem around 1935 (*with discussion*). Statistical Science 1, 78–96.

Lévy, P. (1935). Propriétés asymptotiques des sommes de variables indépendantes on enchainées. *J. Math. Pures Appl.* 14, 347–402.

Lévy, P. (1937). *Théorie de l'Addition des Variables Aléatoires.* Paris: Gauthier-Villars.

Lewin, R. (1978). *Ultra Goes to War.* New York: McGraw-Hill.

Lindeberg, J. W. (1922). Eine neue Herleitung des Exponential-gesetzes in der Wahrscheinlichkeitsrechnung. *Mathematische Zeitschrift* 15, 211–225.

Maistrov, L. E. (1974). *Probability Theory: A Historical Sketch.* New York: Academic Press.

Petrov, V. V. (1975). *Sums of Independent Random Variables.* New York: Springer-Verlag.

Stigler, S. M. (1986). *The History of Statistics.* Cambridge, MA: Harvard University Press.

Turing, A. M. (1934). On the Gaussian error function. Unpublished Fellowship Dissertation, King's College Library, Cambridge.

Welchman, G. (1982). *The Hut Six Story.* New York: McGraw-Hill.

PART THREE

Prediction

10

Predicting the Unpredictable

Abstract. A major difficulty for currently existing theories of inductive inference involves the question of what to do when novel, unknown, or previously unsuspected phenomena occur. In this paper one particular instance of this difficulty is considered, the so-called *sampling of species problem*.

The classical probabilistic theories of inductive inference due to Laplace, Johnson, de Finetti, and Carnap adopt a model of simple enumerative induction in which there are a prespecified number of types or species which may be observed. But, realistically, this is often not the case. In 1838 the English mathematician Augustus De Morgan proposed a modification of the Laplacian model to accommodate situations where the possible types or species to be observed are not assumed to be known in advance; but he did not advance a justification for his solution.

In this paper a general philosophical approach to such problems is suggested, drawing on work of the English mathematician J. F. C. Kingman. It then emerges that the solution advanced by De Morgan has a very deep, if not totally unexpected, justification. The key idea is that although 'exchangeable' random *sequences* are the right objects to consider when all possible outcome-types are known in advance, exchangeable random *partitions* are the right objects to consider when they are not. The result turns out to be very satisfying. The classical theory has several basic elements: a representation theorem for the general exchangeable sequence (the de Finetti representation theorem), a distinguished class of sequences (those employing Dirichlet priors), and a corresponding rule of succession (the continuum of inductive methods). The new theory has parallel basic elements: a representation theorem for the general exchangeable random partition (the Kingman representation theorem), a distinguished class of random partitions (the Poisson-Dirichlet process), and a rule of succession which corresponds to De Morgan's rule.

Reprinted with permission from *Synthese* 90 (1992): 205–232, © 1992 by Kluwer Academic Publishers.

An important question rarely discussed in accounts of inductive inference is what to do when the utterly unexpected occurs, an outcome for which no slot has been provided. Alternatively – since we know this *will* happen on occasion – *how* can we coherently incorporate such new information into the body of our old beliefs? The very attempt to do so seems paradoxical within the framework of Bayesian inference, a theory of consistency between old and new information.

This is not the problem of observing the 'impossible', that is, an event whose possibility we have considered but whose probability we judge to be 0. Rather, the problem arises when we observe an event *whose existence we did not even previously suspect*; this is the so-called problem of 'unanticipated knowledge'. This is a very different problem from the one just mentioned: it is not that we judge such events impossible – indeed, after the fact we may view them as quite plausible – it is just that beforehand we did not even consider their possibility. On the surface there would appear to be no way of incorporating such new information into our system of beliefs, other than starting over from scratch and completely reassessing our subjective probabilities. Coherence of old and new makes no sense here; there are no old beliefs for the new to cohere with.

A special instance of this phenomenon is the so-called *sampling of species problem*. Imagine that we are in a new terrain, and observe the different species present. Based on our past experience, we may anticipate seeing certain old friends – black crows, for example – but stumbling across a giant panda may be a complete surprise. And, yet, all such information will be grist to our mill: if the region is found rich in the variety of species present, the chance of seeing a particular species again may be judged small, while if there are only a few present, the chances of another sighting will be judged quite high. The unanticipated has its uses.

Thus, the problem arises: How can the theory of inductive inference deal with the potential existence of unanticipated knowledge, and, how can such knowledge be rationally incorporated into the corpus of our previous beliefs? How can we predict the occurrence of something we neither know, nor even suspect, exists? Subjective probability and Bayesian inference, despite their many impressive successes, would seem at a loss to handle such a problem given their structure and content. Nevertheless, in 1838 the English mathematician Augustus De Morgan proposed a method for dealing with precisely this difficulty. This paper describes De Morgan's proposal and sets it within the context of other attempts to explain induction in probabilistic terms.

The organization of the paper is as follows. The second section gives some historical background and briefly describes De Morgan's rule. As will be seen, although the statement of the rule is unambiguous, its justification – at least, as described by De Morgan – is unclear, and our goal will be to understand why De Morgan's rule makes sense. We begin this task by briefly reviewing, in the third section of the paper, the classical analysis of the inductive process in probabilistic terms. This is very well-known material, and our goal here is simply to set up a framework in which to place De Morgan's rule. This is then done in the fourth and fifth sections of the paper: the key point is that while 'exchangeable' random *sequences* are the right objects to consider when all possible outcomes are known in advance, exchangeable random *partitions* are the right objects to consider when they are not.

The result turns out to be very satisfying. The classical theory has several basic elements: a representation theorem for the general exchangeable sequence (the 'de Finetti representation theorem'), a distinguished class of sequences (those arising from the so-called 'Dirichlet priors'), a 'rule of succession', specifying the probability of a future outcome (Carnap's 'continuum of inductive methods'), and an urn-model interpretation ('Polya's urn'). The new theory, developed by the English mathematician J. F. C. Kingman for another purpose but ideally suited for this, has parallel basic elements: a representation theorem for the general exchangeable random partition (the Kingman representation theorem), a distinguished class of random partitions (the 'Poisson-Dirichlet process'), an urn-model representation (Hoppe's urn, 1984, sometimes called the 'Chinese restaurant process'), and a rule of succession which corresponds to . . . De Morgan's rule!

The problem considered by De Morgan is closely related to a statistical problem, mentioned earlier, termed 'the sampling of species' problem. There have been a number of attempts to analyze such questions, beginning with the distinguished English statistician R. A. Fisher. This literature is briefly summarized in the final section of the paper, together with some concluding remarks concerning the original inductive problem.

2. THE DE MORGAN PROCESS AND ITS ANTECEDENTS

Hume's problem of induction asks why we expect the future to resemble the past. One of the most common methods of attempting to answer Hume's question invokes probability theory; and *Laplace's rule of succession* is the classical form of this type of explanation. It states that if an event has occurred n times out of N in the past, then the probability that it will occur the next time is $(n + 1)/(N + 2)$. This version of the rule implicitly assumes that possible outcomes are dichotomous; that is, an event of a specified type either did or did not occur. A more complex form of the rule, which can also be found in Laplace's writings, posits instead a multiplicity of possible outcomes. In this setting, the rule becomes: if there are t possible outcomes (labelled $c_1, c_2, \ldots, c_t$), if X_k denotes the outcome occurring on the k-th trial, and if the vector $\boldsymbol{n} = (n_1, n_2, \ldots, n_t)$ records the number of instances in which each of the t possible outcomes occur in a total of N trials, then the probability that an outcome of the j-th type will occur again on the next trial is

LAPLACE'S RULE:

$$P[X_{N+1} = c_j \mid \boldsymbol{n}] = \frac{n_j + 1}{N + t}$$

But as the English mathematician and logician De Morgan noted,

[t]here remains, however, an important case not yet considered; suppose that having obtained t sorts in N drawings, and t sorts only, we do not yet take it for granted that these are all the possible cases, but allow ourselves to imagine there may be sorts not yet come out. (De Morgan 1845, p. 414)

The problem of how to deal with the observation of novel phenomena in Bayesian inference is as old as Bayes's theorem itself. In Price's appendix to Bayes's essay (Bayes 1764, pp. 149–53), Price supposes "a solid or die of whose number of sides and constitution we know nothing; and that we are to judge of these from experiments made in throwing it". Price argues that "the first throw only shows that *it has* the side then thrown", and that it is only "*after* the first throw and not before, [that] we should be in the circumstances required" for the application of Bayes's theorem. Price's subsequent analysis, however, is confined to those cases where our experience is *uniform*, that is, where "the same event has followed without interruption in any one or more subsequent experiments" (e.g., the rising of the sun); or where it is known in advance that there are only two categories (e.g., the drawing of a lottery with *Blanks* and *Prizes*).

Laplace considered the multinomial case where there are three or more categories (Laplace 1781, Section 33), but his analysis is limited to those instances where the number of categories is fixed in advance (but see Hald, 1998, pp. 181–2). De Morgan, in contrast, proposed a simple way of dealing with the possibility of an unknown number of categories (De Morgan 1838, pp. 66–67; 1845, pp. 414–15). If initially there are t possible outcomes known, then De Morgan gives as the probability of seeing the outcome on trial $N + 1$ fall into the j-th category:

DE MORGAN'S RULE:

$$P[X_{N+1} = c_j \mid \boldsymbol{n}] = \frac{n_j + 1}{N + t + 1}.$$

That is, one creates an additional category: "new species not yet observed", which has a probability of $1/(N + t + 1)$ of occurring.

How can one make sense of De Morgan's idea? First, it is unclear what one should do after observing a new 'species'. De Morgan (1845, p. 415) takes t to be the number of species present in the sample at any given instant; so that it increases over time. But if De Morgan's rule is thought of as a generalization of Laplace's, then it is more appropriate to view t as fixed, the number of species known to exist prior to sampling. (This second convention is the one employed below.) Nor is it clear whether De Morgan's prescription is even consistent, in the sense that one can find a probability function on sequences which agrees with his rule. So, the first item of business is to see that this is indeed the case.

2.1. The De Morgan Process

It turns out that there is a simple urn model which generates the sequence of probabilities suggested by De Morgan. Consider an urn with one black ball (the 'mutator'), and t additional balls, each of a different color, say, c_1, $c_2, \ldots, c_t$. We reach into the urn, pick a ball at random, and return it to the urn *together with a new ball*, according to the following rule:

- If a colored ball is drawn, then it is replaced together with another of the *same* color.
- If the mutator is drawn, then it is replaced together with another ball of an *entirely new* color.

The colored balls correspond to species known to exist; selecting a ball of a given color corresponds to observing the species represented by that color; selecting the mutator to observing a hitherto unknown species.

Clearly this sequence of operations generates the probabilities De Morgan suggests. After N drawings, there are $N + t + 1$ balls in the urn, because we started out with t (the colored balls) + 1 (the mutator) and have added N since. Because we are choosing balls at random, each has a probability of 1 in $N + t + 1$ of being selected. The number of colors is gradually changing, but if there are $n_j + 1$ balls of a specific type, then the probability of observing that type at the next draw is the one given by De Morgan. On the other hand, since there is always only one mutator, the probability of it being selected (the probability that a new species is observed) is $1/(N + t + 1)$. This process generates the probabilities specified by De Morgan, so we shall call it the *De Morgan process*.

More generally, we might imagine that the mutator has a 'weight' θ accorded to it, $0 < \theta < \infty$, so that it is either more or less likely to be selected than the colored balls in the urn, which are accorded a weight of 1. That is, each colored ball has a probability of $(N + t + \theta)^{-1}$ of being selected, while the mutator has probability $1 - (N + t)/(N + t + \theta) = \theta/(N + t + \theta)$. This will also be called a De Morgan process (with parameter θ).

So, De Morgan's prescription is consistent. But does it make sense? Isn't it simply arbitrary, no better or worse than any of a broad spectrum of rules we could invent? The answer, surprisingly, is '*No*': it turns out to be a very special process, with many distinctive and attractive features. But, in order to appreciate this, we need to briefly review the classical probabilistic account of induction for a fixed number of categories, and then leap forward nearly a century and a half, when the De Morgan process mysteriously reappears in the 1970s.

3. EXCHANGEABLE RANDOM SEQUENCES

Attempts to explain enumerative induction in probabilistic terms go back to Bayes and Laplace, but this program was perfected at the hands of the twentieth-century Italian mathematician and philosopher Bruno de Finetti. De Finetti's crucial insight was that those situations in which the simplest forms of enumerative induction are appropriate are captured by the mathematical concept of 'exchangeability', and that the mathematical structure of such sequences is readily described.

3.1. The De Finetti Representation Theorem

Let $X_1, X_2, \ldots, X_N, \ldots$ be an infinite sequence of random variables taking on any of a finite number of values, say $c_1, c_2, \ldots, c_t$: these are the possible *categories* or *cells* into which the outcomes of the sequence are classified, and might denote different species in an ecosystem, or words in a language. The sequence is said to be *exchangeable* if for every N the 'cylinder set' probabilities

$$P[X_1 = e_1, X_2 = e_2, \ldots, X_N = e_N] = P[e_1, e_2, \ldots, e_N]$$

are invariant under all possible permutations of the time index. Put another way, two sequences have the same probability if one is a *rearrangement* of

222

the other. If the outcomes are thought of as letters in an alphabet, then this means that all words of the same length having the same letters have the same probability.

Given a sequence of possible outcomes $e_1, e_2, \ldots, e_N$, let n_j denote the number of times the j-th type occurs in the sequence. The frequency vector $\boldsymbol{n} = (n_1, n_2, \ldots, n_t)$ plays a key role in exchangeability (in Carnap's terminology, it is the "structure-description"). First, it provides an equivalent characterization of exchangeability, since given any two sequences, say $\boldsymbol{e} = (e_1, e_2, \ldots, e_N)$ and $\boldsymbol{e}^* = (e_1^*, e_2^*, \ldots, e_N^*)$, one can be obtained from the other by rearrangement if and only if the two have the same frequency vector. Thus, P is exchangeable if and only if two sequences having the same frequency vector have the same probability.

In the language of theoretical statistics, the observed frequency counts $n_j = n_j(X_1, X_2, \ldots, X_N)$ are *sufficient statistics* for the sequence $\{X_1, X_2, \ldots, X_N\}$, in the sense that probabilities conditional on the frequency counts depend only on $\boldsymbol{n}$, and are independent of the choice of exchangeable P: given a particular value of the frequency vector, the only sequences possible are those having this frequency vector, and each of these, by exchangeability, is assumed equally likely. The number of such sequences given by the *multinomial coefficient* $N!/(n_1! n_2! \ldots n_t!)$; and, thus, the probability of such a sequence is

$$P[X_1, X_2, \ldots, X_N \mid \boldsymbol{n}] = \frac{n_1! n_2! \ldots n_t!}{N!}.$$

The structure of exchangeable sequences is actually quite simple. Let

$$\Delta_t =: \{(p_1, p_2, \ldots, p_t): p_i \geq 0 \text{ and } p_1 + p_2 + \cdots + p_t = 1\}$$

denote the t-simplex of probabilities on t elements. Every element of the simplex determines a *multinomial probability*, and the general exchangeable probability is a mixture of these. This is the content of a celebrated theorem due to de Finetti: if an infinite sequence of t-valued random variables $X_1, X_2, X_3, \ldots$ is exchangeable, and $(n_1, n_2, \ldots, n_t)$ is the vector of frequencies for $\{X_1, X_2 \ldots, X_N\}$, then the infinite limiting frequency

$$Z =: \lim_{N \to \infty} \left(\frac{n_1}{N}, \frac{n_2}{N}, \ldots, \frac{n_t}{N} \right)$$

exists almost surely; and, if $\mu(A) = P[Z \in A]$ denotes the distribution of this limiting frequency, then

$$P[X_1 = e_1, X_2 = e_2, \ldots, X_N = e_N]$$
$$= \int_{\Delta_t} \frac{\Gamma(\alpha_1 + \cdots + \alpha_t)}{\Pi \, \Gamma(\alpha_i)} p^{n_1} p^{n_2} \ldots p^{n_t} d\mu(p_1, p_2, \ldots, p_{t-1}).$$

223

The use of such integral representations of course predates de Finetti; de Finetti's contribution was to give a philosophical justification for their use, based on the concept of exchangeability, one not appealing to objective chances or second-order probabilities to explain the nature of the multinomial probabilities appearing in the mixture (e.g., Zabell 1988, 1989).

3.2. Determining the Prior Measure $d\mu$

In order to apply the de Finetti representation theorem, it is necessary to decide on a specific 'prior' $d\mu$. In principle $d\mu$ can be anything, but it is natural to single out classes of priors thought to represent situations of limited knowledge or 'ignorance'. Such ideas go back to Bayes himself, who considered "an event concerning the probability of which we absolutely know nothing antecedently to any trials made concerning it" (Bayes 1764). The earliest and best-known prior is the so-called 'Bayes–Laplace prior', which assumes that there are two categories, say 'success' and 'failure' (so that $t = 2$), and takes $d\mu(p) = dp$. Although Laplace made direct use of this prior, Bayes deduced it by a more circuitous route, assuming that S_N, the number of successes in N trials, is equally likely to assume any value between 0 and N: $P[S_N = k] = 1/(N + 1)$. This assumption in fact uniquely determines $d\mu$ (see Zabell 1988, pp. 159–60).

There is an obvious generalization of Bayes's postulate, employing the frequency vector, which was proposed by the English logician, philosopher, and economic theorist William Ernest Johnson. This is Johnson's "combination postulate" (Johnson 1924): *All ordered t-partitions of N are equally likely.* That is, all possible frequency vectors $n = (n_1, n_2, \ldots, n_t)$ are assumed to have equal probability of occurring. (Note that if $t = 2$, then $(n_1, n_2) = (k, N - k)$ and Johnson's postulate reduces to Bayes's.) Since there are

$$A_{N,t} =: \binom{N + t - 1}{t}$$

ordered t-partitions of N (e.g., Feller 1968, p. 38), each of these, assuming the combination postulate, has probability $1/A_{N,t}$ of occurring. In mathematical probability the frequency counts are often referred to as *occupancy numbers*, and the probability distribution arising from the combination postulate as *Bose–Einstein statistics* (generally, Feller 1968, chapter 2, Section 5). The force of Johnson's combination postulate is that, just as in the binomial case, it uniquely determines the mixing measure $d\mu$; here the uniform or 'flat' prior $d\mu(p_1, p_2, \ldots, p_t) = dp_1 \, dp_2 \ldots dp_{t-1}$, first introduced by Laplace in 1778.

3.3. The Rule of Succession

Once the prior $d\mu$ has been implicitly or explicitly specified, one can immediately calculate the *predictive probabilities* that it gives rise to:

$$P[X_{N+1} = c_i \mid X_1, X_2, \ldots, X_N] = P[X_{N+1} = c_i \mid \boldsymbol{n}].$$

Such a conditional probability is sometimes called a 'rule of succession' (the terminology is due to the English logician John Venn). For example, in the case of the Bayes–Laplace prior (where $t = 2$), a simple integration immediately yields Laplace's rule of succession, $(n_1 + 1)/(N + 1)$; and for Johnson's combination postulate the corresponding rule of succession is $(n_j + 1)/(N + t)$ (Johnson 1924, Appendix). A rule of succession uniquely determines the probability of any possible sequence; and the probability specification on sequences corresponding to the combination postulate is, in Carnap's terminology, the c^* function.

There is an air of arbitrariness about the combination postulate, and both Johnson (and later Carnap) ultimately replaced it with one less stringent, Johnson's 'sufficientness' postulate (the terminology is due to I. J. Good):

$$P[X_{N+1} = i \mid \boldsymbol{n}] = f(n_i, N).$$

That is, the only relevant information conveyed by the sample, vis-à-vis predicting whether the next outcome will fall into a given category, is the number of outcomes observed in that category to date; any knowledge of how outcomes not in that category distribute themselves among the remainder is posited to be irrelevant.

As a consequence of the sufficientness postulate, Johnson was able to derive, just as in the case of the combination postulate, the corresponding rule of succession: if $X_1, X_2, \ldots$ is an exchangeable sequence satisfying the sufficientness postulate, and $t \geq 3$, then (assuming that all cylinder set probabilities are positive so that the relevant conditional probabilities exist)

$$P[X_{N+1} = i \mid \boldsymbol{n}] = \frac{n_i + \alpha}{N + t\alpha}$$

(see, generally, Zabell 1982). The corresponding measure in the de Finetti representation in this case is the *symmetrical Dirichlet prior* with parameter α:

$$d\mu(p_1, p_2, \ldots, p_t) = \frac{\Gamma(t\alpha)}{\Gamma(\alpha)^t} p_1^{\alpha-1} p_2^{\alpha-1} \cdots p_t^{\alpha-1} dp_1 dp_2 \ldots dp_{t-1}.$$

225

3.4. Polya's Urn Model

There is a simple urn model which generates Laplace's rule of succession. It is usually referred to as the Polya urn model (e.g., Feller 1968, pp. 119–21), after the mathematician George Polya, who proposed its use as a model for the spread of contagious diseases, although a description of it (in the case of all successes) can be found in Quetelet's *Lettres sur la théorie des probabilités* (Quetelet 1846, p. 367).

4. PARTITION EXCHANGEABILITY

Johnson's sufficientness postulate, or its later equivalent formulation, Carnap's 'continuum of inductive methods', attempts to capture the concept of prior ignorance about individual categories. Despite its attractiveness, however, it is far from clear that Johnson's sufficientness postulate is a necessary condition for such a state of ignorance. Is it possible to further weaken the notion of absence of information about the categories? A natural idea is that ignorance about individual *categories* should result in a symmetry of beliefs similar to that captured by de Finetti's notion of exchangeability with respect to *times*. This suggests the following definition.

Definition. A probability function P is *partition exchangeable* if the cylinder set probabilities $P[X_1 = e_1, X_2 = e_2, \ldots, X_N = e_N]$ are invariant under permutations of the time index *and* the category index.

For example, if we are rolling a die (so that $t = 6$), and our subjective probabilities for the various outcomes are partition exchangeable, then

$$P[6, 4, 6, 4, 4, 5, 1, 2, 5] = P[1, 1, 1, 2, 2, 3, 3, 4, 5].$$

This can be seen by first arranging the sequence

$$\{6, 4, 6, 4, 4, 5, 1, 2, 5\}$$

into 'regular position':

$$\{4, 4, 4, 5, 5, 6, 6, 1, 2\},$$

(i.e., descending order of observed frequency for each face); and then follow this up by the category permutation

$$1 \to 4 \to 1, 2 \to 5 \to 2, 3 \to 6 \to 3,$$

which can be more compactly written as $(1, 4)(2, 5)(3, 6)$.

226

The 'sufficient statistics' for a partition exchangeable sequence are the *frequencies of the frequencies* (or '*abundances*'):

$$a_r =: \text{number of } n_j \text{ equal to } r$$

Example. Suppose one observes the sequence 5, 2, 6, 1, 2, 3, 5, 1, 1, 2. Then:

$$N = 10; t = 6.$$
$$n_1 = 3, n_2 = 3, n_3 = 1, n_4 = 0, n_5 = 2, n_6 = 1.$$
$$\boldsymbol{n} = (3, 3, 1, 0, 2, 1) \text{ "} = \text{" } 0^1 1^2 2^1 3^2$$
$$a_0 = 1, a_1 = 2, a_2 = 1, a_3 = 2, a_4 = \cdots a_{10} = 0.$$
$$a = (1, 2, 1, 2, 0, \ldots, 0)$$

A useful bit of terminology will be to call the $\boldsymbol{a}$-vector the *partition vector*. (Kingman (1980, p. 36) calls it the "allelic partition".) Note that in a partition exchangeable sequence, $P[X_1 = 1/t]$, so the number of categories that appear in such a sequence must be finite.

The partition vector plays the same role relative to partition exchangeable sequences that the frequency vector plays for ordinary exchangeable sequences; that is, two sequences are equivalent, in the sense that one can be obtained from the other by a permutation of the time set and a permutation of the category set, if and only if the two sequences have the same partition vector. Thus, an alternative characterization of partition exchangeability is that: *all sequences having the same partition vector have the same probability.* The frequencies of the frequencies, furthermore, are the sufficient statistics for a partition exchangeable sequence, since probabilities conditional on the partition vector $\boldsymbol{a} = (a_0, a_1, \ldots, a_t)$ are independent of P: and, given a partition vector $\boldsymbol{a}$, the only possible sequences have $\boldsymbol{a}$ as partition vector and each of these is equally likely. (Note that this refers only to the cylinder set probabilities involving $X_1, X_2, \ldots, X_N$. The predictive probabilities for $X_{N+1}, X_{N+2}, \ldots$ will still depend on the a_r.)

According to the de Finetti representation theorem, a partition exchangeable sequence, being exchangeable, can be represented by a mixing measure $d\mu$ on the t-simplex Δ_t. An important subset of the t-simplex in the partition exchangeable case is the subsimplex of ordered probabilities:

$$\Delta_t^* = \{(p_1^*, p_2^*, \ldots, p_t^*) : p_1^* \geq p_2^* \geq \cdots \geq p_t^* \geq 0, \Sigma_j p_j^* = 1\}$$

In the partition exchangeable case, once the prior $d\mu$ is known on the ordered t-simplex Δ_t^*, it is automatically determined on all of Δ_t by symmetry.

It is not really difficult to prove this, but it is perhaps best seen by considering a few simple examples.

Consider, first, the case of a coin which we know to be biased 2:1 in favor of one side, but where we don't know which side it is – it could be either with equal probability. Then, $p_1^* = 2/3$, $p_2^* = 1/3$. In terms of the original, unordered probabilities, this corresponds to either $p_1 = 2/3$, $p_2 = 1/3$ or $p_1 = 1/3$, $p_2 = 2/3$ and, since we are indifferent between categories, these two possibilities are equally likely; thus, we have as the mixing measure on the simplex Δ_2 the measure on the first component $p = p_1$

$$d\mu(p) = \frac{1}{2}\delta_{2/3} + \frac{1}{2}\delta_{1/3},$$

where δ_x is the Dirac measure which assigns probability 1 to x. This is a partition exchangeable probability, since it is invariant under the interchange $H \to T, T \to H$.

Consider next a die with six faces. The most general exchangeable probability is obtained by mixing multinomial probabilities over the simplex Δ_6. The partition exchangeable probabilities are those which are invariant with respect to interchange of the faces. This would be equivalent to specifying a probability over

$$\Delta_6^* = \left\{(p_1^*, p_2^*, \ldots, p_6^*): p_1^* \geq \cdots \geq p_6^* \geq 0, \sum_{j=1}^{6} p_j^* = 1\right\}.$$

Specifying such a probability would be to say we have opinions about the bias of the die (the most likely face has probability p_1^*, the second most likely p_2^*, and so on), but not about *which* face has the bias, since our probability function is symmetric with respect to faces.

A little thought should make it clear that the frequencies of frequencies can provide information relevant to the prior $d\mu$ on Δ_t^* (in the partition exchangeable case). For example, suppose that we know the die is biased in favor of one face, and that the other faces are equally likely. Then, the unknown vector of ordered probabilities satisfies $p_1^* > p_2^* = p_3^* = \cdots = p_6^*$. Suppose now that in 100 trials we observe the frequency vector (20, 16, 16, 16, 16, 16). Then, we would guess that $p_1 = p_1^* = .2$ (approximately), and $p_2 = p_2^* = p_3 \cdots = p_6^* = .16$. But, if we observed the frequency vector (20, 40, 10, 15, 10, 5), we would guess $p_2 = p_1^* = .4$, and $p_1 = p_2^* = (20 + 10 + 15 + 10 + 5)/\{(100)(5)\} = .12$. Our estimate for p_1 differs in the two cases (.16 vs. .12) despite the fact that the frequency count for the first category is the same in both cases.

This is clearly, then, an objection to Johnson's sufficientness postulate (and, thus, also Carnap's continuum of inductive methods): although on the surface it appears to be a reasonable quantification of a state of ignorance about individual categories, it asserts that the frequencies of the frequencies lack relevant information about the probabilities of those categories. Nevertheless, as the example demonstrates, it is certainly possible to have degrees of belief which are category symmetric, and yet for which the frequencies of frequencies provide very real information. This far from obvious fact was apparently first noted by the brilliant English mathematician Alan Turing during World War II (see Good 1965, p. 68; 1979).

In general, the predictive probabilities for partition exchangeable probabilities will have the form

$$P[X_{N+1} = c_i | X_1, X_2, \ldots, X_N] = f(n_i, a_0, a_1, \ldots, a_N).$$

Johnson's sufficientness postulate thus makes the very strong supposition that the predictive probabilities reduce to a function $f(n_i, N)$. In a very interesting paper, Hintikka and Niiniluoto (1980) explore the consequences of the weaker assumption that the predictive probabilities are functions $f(n_i, a_0, N)$; that is, these may also depend on the number of categories which are thus far unobserved. This generalization of Johnson's postulate seems very natural within the context of partition exchangeability, but it is unclear why the dependence on the partition vector should be limited to only its first component. Ultimately it is only partition exchangeability which exactly captures the notion of complete ignorance about categories; any further restriction on a probability beyond that of category symmetry necessarily involves, at least implicitly, some assumption about the categories. The temptation to do so, of course, is understandable; unlike the continuum of inductive methods, the partition exchangeable probabilities do not form a finite-dimensional family, which can be described by a finite number of parameters.

NOTE: In general there are $t!$ permutations of the set of integers $\{1, 2, \ldots, t\}$; and to every such permutation there corresponds a subsimplex $\Delta_{t,\sigma}$ of Δ_t, namely, $\Delta_{t,\sigma} = \{(p_1, p_2, \ldots, p_t) \in \Delta_t : p_{\sigma(1)} \geq p_{\sigma(2)} \geq \cdots \geq p_{\sigma(t)}\}$. The map $(p_1, p_2, \ldots, p_t) \to (p_{\sigma(1)}, p_{\sigma(2)}, \ldots, p_{\sigma(t)})$ defines a homeomorphism of $\Delta_{t,\sigma}$ onto Δ_t^*, and this map permits one to transfer the values of a prior $d\mu$ on Δ_t^* to the subsimplex $\Delta_{t,\sigma}$.

5. EXCHANGEABLE RANDOM PARTITIONS

Now we come to the major point of this paper. How can a Bayesian allow for (1) infinite categories, or (2) unknown species?

If the number of categories is infinite, then no prior can be category symmetric, for such a prior would have to assign equal weight to each category, which is impossible; that is, if there are an infinite number of colors (say), $c_1, c_2, \ldots$, then $P[X_1 = c_1] = P[X_2 = c_2] = \cdots 1/t$, which is impossible, since $t = \infty$. We are thus compelled to consider probability assignments which contain some element of asymmetry between the different categories.

But, more seriously, *what* does it even mean to assign probabilities in a situation where we are encountering previously unknown species, continuously observing new and possibly unsuspected kinds? According to (at least one naive version of) the classical Bayesian picture, one assigns probabilities in advance to all possible outcomes and, then, updates via Bayes's theorem as new information comes in. How can one introspect and assign probabilities when the possible outcomes are unknown beforehand?

The earlier discussion of partition exchangeable sequences suggests a solution to this second difficulty: rather than focus on the probability of a sequence of outcomes $(e_1, e_2, \ldots, e_N)$, or the probability of a frequency vector $(n_1, n_2, \ldots, n_t)$ (the elements of which refer to specific species), focus instead on the partition vector $(a_1, a_2, \ldots, a_N)$ and its probability. Even if one does not know *which* species are present prior to sampling, one can still have beliefs as to the *relative abundances* in which those species, as yet unobserved, will occur. (Note that in this setting a_0 is excluded from the partition vector: lacking prior knowledge as to the totality of species present, it is impossible to specify at any given stage how many species present do not yet appear in the sample.)

One could, in fact, now proceed exclusively at the level of partition vectors, and construct a theory of the type we are seeking (although it is far from obvious at this stage how to cope in a category symmetric fashion with the $t = \infty$ case discussed above). But there would appear to be a substantial cost: the rich theoretical structure of exchangeability, the representation theorem, ignorance priors, and the like. One need not despair, however. All this can be obtained, *provided* one looks at the matter in a new, if initially somewhat unorthodox, manner.

5.1. Exchangeable Random Partitions

The key point is to recognize that in the sampling of species scenario, the relevant information being received is an *exchangeable random partition*. Because the individual species do not, in effect, have an individuality – we simply observe the first species, then at some subsequent time the second, at a still later time the third, and so on – the relevant information being received is a *partition* of the integers.

In other words, the first species occurs at some set of times

$$A_1 =: \left\{ t_1^1, t_1^2, t_1^3, \ldots : t_1^1 < t_1^2 < t_1^3 < \cdots \right\}$$

where necessarily $t_1^1 = 1$, and in general the set A_1 may only contain a finite number of times even if an infinite number of observations is made (this will happen if the first species is only observed a finite number of times, possibly even only once, in which case $A_1 = \{t_1^1\}$). Likewise, the second species occurs at some set of times

$$A_2 =: \left\{ t_2^1, t_2^2, t_2^3, \ldots : t_2^1 < t_2^2 < t_2^3 < \cdots \right\}$$

where necessarily t_2^1 is the first positive integer not in A_1, and A_2 may again be finite. In general, the i-th species to be observed occurs at some set of times $A_i = \{t_i^j : j = 1, 2, 3, \ldots\}$ and the collection of sets $A_1, A_2, A_3, \ldots$ forms a *partition* of the positive integers N in the sense that

$$N = A_1 \cup A_2 \cup A_3 \cup \cdots \quad \text{and} \quad A_i \cap A_j = \emptyset, i \neq j.$$

In the example considered before, the partition of $\{1, 2, 3, \ldots 10\}$ generated is

$$\{1, 7\} \cup \{2, 5, 10\} \cup \{3\} \cup \{4, 8, 9\} \cup \{6\}.$$

Note another interpretation we can now give the partition vector $a = (a_1, a_2, \ldots, a_{10})$: it records the sizes of the sets in the partition and the number of species observed. Thus, in our example, given the partition vector is $(2, 1, 2, 0, \ldots, 0)$, two sets in the resulting partition have a single element (since $a_1 = 2$), one set in the partition has two elements (since $a_2 = 1$), two sets in the partition have three elements (since $a_3 = 2$), and the total number of species observed is 5 (since $a_1 + a_2 + \cdots + a_{10} = 5$). Although originally defined in terms of the underlying sequence, the partition vector is a function solely of the resulting partition of the time set; and one can therefore refer to the partition vector of a partition.

Thus, observing the successive species in our sample generates a random partition of the positive integers. Now let us consider in what sense such a partition could be 'exchangeable'. An obvious idea is to examine the structure of random partitions arising from exchangeable sequences and see if we can characterize them in some way.

This turns out to be relatively simple: *if a random sequence is exchangeable, then the partition structures for two possible sequences have the same probability whenever they have the same partition vector **a**.* In order to see this, let's think about what happens to a partition when we permute the categories or times of the underlying sequence which generates it. Consider our earlier example of the sequence $\{5, 2, 6, 1, 2, 3, 5, 1, 1, 2\}$, and suppose we permute the category index in some way, say, the cyclic permutation

$$1 \to 2 \to 3 \to 4 \to 5 \to 6 \to 1.$$

Then, our original sequence becomes transformed into $\{6, 3, 1, 2, 3, 4, 6, 2, 2, 3\}$, and the resulting partition of the time set from 1 to 10 is the same as before: species 6 occurs at times 1 and 7, hence, we get $A_1 = \{1, 7\}$, and so on. *Permuting the category index results in a new sequence but leaves the resulting partition unchanged.*

Next, suppose we were to permute the times, say, by the cyclic permutation

$$1 \to 2 \to 3 \to 4 \to 5 \to 6 \to 7 \to 8 \to 9 \to 10 \to 1.$$

(That is, what happened at time 1 is observed to occur at time 2 instead; at time 2, at time 3 instead; and so on.)

Then, our original sequence becomes transformed into $\{2, 5, 2, 6, 1, 2, 3, 5, 1, 1\}$, and we get a new partition of the time set, namely,

$$\{1, 3, 6\} \cup \{2, 8\} \cup \{4\} \cup \{5, 9, 10\} \cup \{7\}.$$

Because of the exchangeability of the underlying sequence, this new partition has the same probability of occurring as the original one. Note that it has the same frequency vector **n** and, therefore, partition vector **a**. This observation is the one underlying the idea of an exchangeable random partition:

Definition. A random partition is *exchangeable* if any two partitions π_1 and π_2 having the same partition vector have the same probability; i.e., if

$$a(\pi_1) = a(\pi_2) \Rightarrow P[\pi_1] = P[\pi_2].$$

5.2. The Kingman Representation Theorem

In the case of sequences, the de Finetti representation theorem states that the general exchangeable sequence can be constructed out of elementary building blocks: Bernoulli trials (coin-tossing sequences) in the case of 0,1-valued random variables; multinomial trials in the case of t-valued random variables;

and in general sequences of independent and identically-distributed random variables. The corresponding building blocks of the general exchangeable random partition are the *paintbox processes*.

In order to construct a paintbox process, consider an ordered 'defective' probability vector

$$p = (p_1, p_2, p_3, \ldots), \quad \text{where} \quad p_1 \geq p_2 \geq p_3 \cdots \geq 0$$
$$\text{and} \quad p_1 + p_2 + p_3 + \cdots \leq 1,$$

and let ∇ denote the infinite simplex of all such vectors.

Given such a defective probability vector $p = (p_1, p_2, p_3, \ldots) \in \nabla$, let $p_0 =: 1 - \Sigma_i p_i$; and let μ_p be a probability measure on the unit interval $[0,1]$ having point masses p_j at some set of distinct points x_j, $j \geq 1$ (which points are selected doesn't matter), and a continuous component assigning mass p_0 to $[0, 1]$. Call such a probability measure a *representing probability measure* for p. Let $X_1, X_2, X_3, \ldots$ be a sequence of independent and identically-distributed random variables with common distribution μ_p, and consider the exchangeable random partition generated by the rule:

$$A_j = \{i : X_i = x_j\} \quad \text{where} \quad A_1 \cup A_2 \cup \cdots = \{1, 2, \ldots, N\}.$$

That is, partition the integers $1, 2, \ldots, N$ by grouping together those times i at which the random variables X_i have a common value x_j.

It is then not difficult to see that if $p \in \nabla$, and μ_p and ν_p are two different representing probability measures for p, then μ_p and ν_p generate the same exchangeable random partition Π, in the sense that the two random partitions have the same stochastic structure. Thus, we have a well-defined rule for associating exchangeable random partitions with vectors in ∇: given p, select μ_p, and use μ_p to generate Π. Let's call this resulting exchangeable random partition Π_p. This is a paintbox process.

Now we are ready to state the Kingman representation theorem:

Theorem (Kingman 1978). *The general exchangeable random partition is a mixture of paintbox processes.*

Let us make this precise. Suppose that $Z_1, Z_2, Z_3, \ldots$ is a sequence of random partitions; specifically, for each $N \geq 1$, Z_N is an exchangeable random partition of $\{1, 2, \ldots, N\}$. There is an obvious sense in which such a sequence is consistent. Namely, any partition of $(1, 2, \ldots, N + 1)$ gives rise to a partition of $\{1, 2, \ldots, N\}$ by simply omitting the integer $N + 1$ from the subset in which it occurs. Let $T_{N+1,N}$ denote the map which performs this operation. Then, the pair Z_{N+1} and Z_N are *consistent* if $P[Z_N \in A] = P[T_{N+1,N}(Z_{N+1}) \in A]$,

233

where A is a set of partitions of $\{1, 2, \ldots, N\}$; and the sequence is consistent if Z_N and Z_{N+1} are consistent for every $N \geq 1$. Every such consistent sequence gives rise to a probability measure on the partitions of $N =: \{1, 2, 3, \ldots\}$. If Π is the probability distribution on the partitions of N arising from such an arbitrary exchangeable random partition, then the Kingman representation theorem states that there exists a (unique) probability measure $d\mu$ on ∇, the infinite simplex of all ordered defective probability vectors, such that for every (measurable) set A of partitions,

$$\Pi(A) = \int_{\nabla} \Pi_P(A) \, d\mu(\boldsymbol{p}).$$

Note that instead of integrating over the probability simplex, one integrates over the ordered defective probability simplex ∇ consisting of all possible defective probability vectors $\boldsymbol{p}$. Moreover, as proven by Kingman, the ordered sample frequencies arising from the random partition converge in joint distribution to the mixing measure $d\mu$. (Just as in de Finetti's theorem the *unordered* sample frequencies $(n_1/N, \ldots, n_t/N)$ converge to the mixing measure $d\mu$, in the de Finetti representation, here the *ordered* sample frequencies converge to the mixing measure $d\mu$ in the Kingman representation.)

The distinctive role that the *continuous* component p_0 of a paintbox process plays in the theorem deserves some comment. When Kingman first investigated exchangeable random partitions, he was puzzled by the fact that mixtures over the *discrete* nondefective ordered probabilities $(p_1^*, p_2^*, p_3^*, \ldots)$ generated many, but by no means all possible exchangeable random partitions. The key to this puzzle is the far from obvious observation that when a new species appears, it must always suffer one of two fates: either it never appears again, or it is subsequently seen an infinite number of times. No intermediate fate is possible. The species that arise once and only once are precisely those that arise from the continuous component.

The Reverend Dr. Richard Price would not have found this surprising. As he states (Bayes 1764, p. 312), the first appearance of an event only informs us of its *possibility*, but would not "give us the least reason to apprehend that it was, in that instance or in any other, regular rather than irregular in its operations"; that is, we are given no reason to think that its probability of recurring is positive (read "regular") rather than 0 (read "irregular"). In effect, Price is saying that the first observation tells us that the outcome lies in the *support* of the unknown representing probability μ_p, while the second observation tells us that it lies in the *discrete component* of this probability.

234

5.3. The Poisson–Dirichlet Process

Thus far we have managed to capture a notion of exchangeable random outcome suitable to the sampling of species setting, and have a representation theorem as well. But the classical theories of induction that employ probability theory usually attempt to go further and identify classes of possible priors $d\mu$ thought to represent situations of limited information. In the de Finetti representation discussed earlier, this was easy: the so-called flat priors dp or $dp_1 dp_2 \dots dp_{t-1}$ immediately suggested themselves, and the game was to come up with characterizations of these priors in terms of symmetry assumptions about the underlying cylinder set probabilities. Here, however, it is far from apparent what a 'flat' prior would be.

At this point we encounter a deep and truly ingenious idea of Kingman's. Let $\alpha > 0$. Suppose we took a symmetric Dirichlet prior $\boldsymbol{D}(\alpha)$ on the t-simplex Δ_t and let the number of categories tend to infinity (i.e., let $t \to \infty$). The resulting probabilities would then 'wash out'. For any fixed $t_0 < \infty$ and $(x_1, x_2, \dots x_{t_0}) \in \Delta_t$, the cylinder set probabilities

$$P_{\alpha,t}[p_1 \leq x_1, p_2 \leq x_2, \dots, p_{t_0} \leq x_{t_0}] \to 0 \quad \text{as } t \to \infty.$$

But, suppose instead that we consider the vector of ordered probabilities. Then, something truly remarkable occurs. Since we can map the t-simplex Δ_t onto the ordered t-simplex Δ_t^* (by associating to any vector $(p_1, p_2, \dots, p_t)$ its ordered rearrangement $(p_1^*, p_2^*, \dots, p_t^*)$), the symmetric Dirichlet prior on Δ_t induces a probability distribution on Δ_t^*: for any fixed $t_0 \leq t < \infty$ and sequence $(x_1^*, x_2^*, \dots, x_{t_0}^*) \in \Delta_t^*$, there is a corresponding cylinder set probability

$$P_{\alpha,t}[p_1^* \leq x_1^*, p_2^* \leq x_2^*, \dots, p_{t_0}^* \leq x_{t_0}^*].$$

Then, as Kingman shows, if $t \to \infty$ and $\alpha \to 0$ in such a way that $t\alpha \to \theta > 0$, for some positive number θ, then the resulting sequence of probabilities does not 'wash out': instead, it has a proper limiting distribution. And, since this is so for each t, the result is a probability measure on ∇. (A 'consistent' set of probabilities on the finite cylinder sets always corresponds to a unique probability on infinite sequence space.) This is called the *Poisson–Dirichlet distribution* (with parameter θ). (The terminology is intended to suggest an analogy with the classical Poisson-binomial limit theorem in probability theory.)

A simple example will illustrate the phenomenon. Suppose you pick a point $\boldsymbol{p}$ at random from Δ_t according the symmetric Dirichlet distribution $P_{\alpha,t}$ and ask for the probability $P_{\alpha,t}[p_1 \geq x_1]$. As $t \to \infty$, this probability tends to 0 (since a typical coordinate of $\boldsymbol{p}$ will be small if t is large). But suppose, instead, you ask for the probability that the *maximum* coordinate of p exceeds x_1: that is, $P_{\alpha,t}[p_1^* \geq x_1]$. Then, Kingman's theorem states that this probability has a *nonzero* limit as $t \to \infty$. Such a result, although hardly obvious, is evidently neither counterintuitive nor paradoxical.

5.4. The Ewens Sampling Formula

Since the Poisson–Dirichlet distribution with parameter θ is a probability measure on ∇, and each paintbox process in ∇ gives rise to an exchangeable random partition, for every sample size n the Poisson–Dirichlet distribution induces a probability distribution $P[a_1, a_2, \ldots, a_n]$ on the set of possible partition vectors. Kingman shows that these probabilities are given by the so-called

EWENS SAMPLING FORMULA:

$$\frac{n!}{\theta(\theta+1)\cdots(\theta+n-1)} \prod_{r=1}^{n} \frac{\theta^{a_r}}{r^{a_r} a_r!}$$

This little formula turns out to be remarkably ubiquitous: it is called the Ewens sampling formula, because it was first discovered by the geneticist Warren Ewens in the course of his work in theoretical population genetics (Ewens 1972). It crops up in a large number of seemingly unrelated contexts. One example of many is: if one picks a random permutation of the integers $\{1, 2, \ldots, N\}$, and lets a_j denote the number of j-cycles, then the probability distribution for $a_1, a_2, \ldots, a_N$ is provided by the Ewens formula.

Given the Ewens formula for the cylinder set probabilities, it is a simple calculation to derive the corresponding predictive probabilities or rules of succession. It is important, however, to be clear what this means, so let's back up for a moment. Suppose we are performing a sequence of observations X_1, $X_2, \ldots, X_N, \ldots$, noting at each stage either the species of an animal, the next word used by Shakespeare, or whatever. At each point, we observe either a species previously observed or an entirely new species. Before these are observed, it doesn't make sense to refer to these outcomes as exchangeable; in fact, it doesn't even make sense to refer to the probabilities of such outcomes, because ahead of time we don't know what a complete list of possible outcomes is. We're learning as we go along. But at time N we can construct a partition of $\{1, 2, \ldots, N\}$ on the basis of what we've seen thus far, and it *does* make sense to talk prospectively about the probability of seeing a particular partition. It is then natural to assume that the resulting random partition is exchangeable; it is necessary to tutor one's intuition, but this is the end result.

236

(As Diaconis and Freedman (1980, p. 248) observe about the concept of Markov exchangeability, "the notion of symmetry seems strange at first. . . . A feeling of naturalness only appears after experience and reflection".) Having arrived at this epistemic state, we can then invoke the Kingman representation theorem, and write our exchangeable random partition as a mixture of paint-box processes. Although we do not, indeed cannot, have prior beliefs about the probabilities of the species we observe, since we didn't know they existed until we saw them, we can certainly have opinions about their *abundances*: that is, what is the frequency of occurrence of the most abundant species, the second most abundant, and so on, and this is what our prior on ∇ summarizes.

Now, given that we make a series of N observations, it is clear that our exchangeable probability assignment will predict whether a new species will be observed on the next trial. And, if we don't observe a new species, whether we see a member of the same species as the very first animal observed. (That is, whether the new partition resulting after time $N + 1$ will add the integer $N + 1$ to the member of the partition containing 1.) Or, whether a member of the second species observed. (That is, whether the new partition adds $N + 1$ to that member of the partition containing the first integer not in the member of the partition containing 1.) And so on.

Given that we have observed a number of species so far – with n_1 of the first type, n_2 of the second, and so on – *what* are the resulting succession probabilities for observing one of the known species or an unknown one? The answer, given the Poisson–Dirichlet prior (and letting s_j denote the j-th species observed to date) is:

$$P[X_{N+1} = s_j \mid n] = \frac{n_j}{(N + \theta)}$$

That is, with $\theta = 1$ and $t = 0$ *the answer is identical to De Morgan's!*

Thus, De Morgan's answer emerges as far from arbitrary. It arises from the canonical 'ignorance prior' for exchangeable random partitions.

5.5. The Chinese Restaurant Process

Completing our analogy with the case of exchangeable sequences, what is the generating urn process for this 'benchmark' process? We already know the answer to this: it is "Hoppe's urn" (Hoppe, 1984), a classical urn model with the added facet of a black ball representing the 'mutator'.

This process has in fact been independently noted several times during the last two decades. Perhaps the most attractive version is the *Chinese restaurant process*: on any given evening in Berkeley, a large number of people go to some Chinese restaurant in the downtown area. As each person arrives, he looks in the window of each restaurant to decide whether or not to go inside. His chance of going in increases with the number of people already seen to be inside, since he takes that as a good sign. But there's always some probability that he goes to an empty restaurant. In a second (and, in fact, the original) version of the process, people enter a single restaurant and sit down at random at one of several circular tables (see Aldous 1985, p. 92). (The main point of this version is that the groups around the tables define the cycles of a random permutation.)

6. SOME FURTHER LITERATURE

The problem discussed above is often referred to in the statistical literature as the *sampling of species problem*. One of the earliest references is a short but important paper by Fisher (Fisher et al., 1943). The sampling of species problem has since been considered by several people from a Bayesian perspective. As noted earlier, Turing seems to have been the first to realize the potential informativeness of the frequencies of the frequencies, a discovery he made during the course of his cryptanalytic work at Bletchley Park during World War II. The noted Bayesian statistician I. J. Good was Turing's statistical assistant at the time, and after the war he published a series of interesting papers in this area (e.g., Good 1953; Good and Toulmin 1956; and Good 1965, chapter 8). These methods have recently been employed to estimate the total number of words known to Shakespeare (Efron and Thisted 1976), and to test whether a poem attributed to Shakespeare was in fact authored by him (Thisted and Efron 1987). During the last two decades the American statistician Bruce Hill has also investigated the sampling of species problem (e.g., Hill 1968, 1979). *Zipf's law* is an empirical relationship that the elements of a partition vector are often found to follow (see Hill 1970). Hill (1988) discusses some relationships between his own methods and those of Kingman.

Kingman's beautiful work is summarized in his monograph, *The Mathematics of Genetic Diversity* (1980; see also Kingman 1975). Kingman's theory was originally stated in terms of "partition structures" (Kingman 1978a), as was his original proof of the representation theorem for exchangeable random partitions (Kingman 1978b). The account given above draws heavily on Aldous (1985, pp. 85–92). The Ewens sampling formula was of course discovered by Ewens (1972); it thus provides a counterexample to Stigler's law of eponomy, but it was also independently discovered shortly after by Charles Antoniak in a Bayesian setting (Antoniak 1974). The urn model discussed in Section 2 is implicit in De Morgan (1838, 1845), but was never formally stated by him. During the 1970s the model surfaced in Berkeley, first as a special case of a class of urn models discussed by Blackwell and MacQueen (1973) and, then, in the guise of the Chinese restaurant process (fathered by Lester Dubins and Jim Pitman). The CRP remained 'folklore', however, until it was described in Aldous's 1985 monograph. The urn model itself first appeared in print in 1984, when Fred Hoppe drew attention to it as a simple method of generating the Ewens sampling formula (see Hoppe 1984, 1987; and Donnelly 1986).

An axiom corresponding to the assumption of partition exchangeability is briefly mentioned by Carnap at the beginning of his book (Carnap 1950), but not pursued further by him. Good has studied priors for multinomial probabilities which are mixtures of symmetric Dirichlet priors (and therefore partition exchangeable); there is a close relationship between some of his work (Good 1953) and recent efforts by Theo Kuipers (1986) to estimate the λ-parameter in Carnap's continuum of inductive methods (equivalently, the α-parameter of the corresponding symmetric Dirichlet prior). Kuipers had earlier discussed a mutation model similar to De Morgan's, but in his system the mutation rate does not tend to zero (see Kuipers 1973).

The concept of exchangeability was introduced into the philosophical literature by Johnson, who termed it the "permutation postulate", and analyzed its consequences assuming first the combination postulate (Johnson 1924) and then the less restrictive sufficientness postulate (Johnson 1932). Exchangeability was soon after independently discovered by de Finetti, who skillfully employed his representation theorem to analyze the structure of the general exchangeable sequence, making no appeal to additional, restrictive postulates. After World War II, Carnap investigated exchangeability as part of a broad attack on the problem of inductive inference, rediscovering many of Johnson's results and carrying his investigations into new territory (see, especially, Carnap 1980).

It is an important historical footnote that Carnap clearly recognized the importance of studying the case of inductive inference when the number of categories is not fixed in advance, and thought that this could be done by employing the equivalence relation R: *belongs to the same species as*. (That is, one has a notion of equivalence or common membership in a species, without prior knowledge of that species.) Carnap did not pursue this idea any further, however, because he judged that it would introduce further complexities into the analysis, which would have been premature given the relatively primitive state of the subject at that time. (My thanks to Richard Jeffrey, to whom I owe the information in this paragraph.)

As we can now appreciate, Carnap displayed great prescience here: the use of such an equivalence relation would have been tantamount to considering partitions rather than sequences, and the resulting complexities are indeed an order of magnitude greater. That we can now see further today is a tribute to the beautiful and profound work of Kingman discussed above.

ACKNOWLEDGMENT

I thank Domenico Costantini, Persi Diaconis, Ubaldo Garibaldi, Tom Nagylaki, and Jim Pitman for helpful discussions and references, and Richard Jeffrey for his comments on a draft of the paper.

REFERENCES

Aldous, D. J. 1985. 'Exchangeability and Related Topics', in P. L. Hennequin (ed.), *École d'Été de Probabilités de Saint-Flour XIII – 1983, Lecture Notes in Mathematics* **1117**, 1–198.

Antoniak, C. E. 1974. 'Mixtures of Dirichlet Processes with Applications to Bayesian Nonparametric Problems', *Annals of Statistics* **2**, 1152–74.

Bayes, T. 1764. 'An Essay Towards Solving a Problem in the Doctrine of Chances', *Philosophical Transactions of the Royal Society of London* **53**, 370–418 (reprinted: 1958, *Biometrika* **45**, 293–315 (page citations in the text are to this edition)).

Blackwell, D. and MacQueen, J. B. 1973. 'Ferguson Distributions via Polya Urn Schemes', *Annals of Statistics* **1**, 353–55.

Carnap, Rudolph. 1950. *Logical Foundations of Probability*, Chicago: University of Chicago Press.

Carnap, R. 1980. 'A Basic System of Inductive Logic, Part II', in R. C. Jeffrey (ed.), *Studies in Inductive Logic and Probability*, Vol. 2, Berkeley and Los Angeles: University of California Press, pp. 7–155.

De Finetti, B. 1937. 'La prevision: ses lois logiques, ses sources subjectives', *Annales de l'Institut Henri Poincaré* **7**, 1–68.

De Morgan, Augustus 1838. *An Essay on Probabilities, and on their Application to Life Contingencies and Insurance Offices*, London: Longman et al.

De Morgan, A. 1845. 'Theory of Probabilities', in *Encyclopedia Metropolitana, Volume 2: Pure Mathematics*, London: B. Fellowes et al., pp. 393–490.

Diaconis, P. and Freedman, D. 1980. 'De Finetti's Generalizations of Exchangeability', in R. C. Jeffrey (ed.), *Studies in Inductive Logic and Probability*, Vol. 2, Berkeley and Los Angeles: University of California Press, pp. 233–50.

Donnelly, P. 1986. 'Partition Structures, Polya Urns, the Ewens Sampling Formula, and the Ages of Alleles', *Theoretical Population Biology* **30**, 271–88.

Efron, B. and Thisted, R. 1976. 'Estimating the Number of Unseen Species: How Many Words did Shakespeare Know?', *Biometrika* **63**, 435–47.

Ewens, W. J. 1972. 'The Sampling Theory of Selectively Neutral Alleles', *Theoretical Population Biology* **3**, 87–112.

Feller, William 1968. *An Introduction to Probability Theory and its Applications*, Vol. 1, 3rd ed., New York: Wiley.

Fisher, R. A., Corbet, A. S. and Williams, C. B. 1943. 'The Relation Between the Number of Species and the Number of Individuals in a Random Sample of an Animal Population', *Journal of Animal Ecology* **12**, 42–58.

Good, I. J. 1953. 'On the Population Frequencies of Species and the Estimation of Population Parameters', *Biometrika* **40**, 237–64.

Good, I. J. 1965. *The Estimation of Probabilities: An Essay on Modern Bayesian Methods*, Cambridge MA: M.I.T. Press.

Good, I. J. 1979. 'Turing's Statistical Work in World War II', *Biometrika* **66**, 393–96.

Good, I. J. and Toulmin, G. H. 1956. 'The Number of New Species, and the Increase in Population Coverage, When a Sample is Increased', *Biometrika* **43**, 45–63.

Hald, A. 1998. *A History of Mathematical Probability and Statistics from 1750 to 1930*, New York: Wiley.

Hill, B. 1968. 'Posterior Distribution of Percentiles: Bayes's Theorem for Sampling from a Finite Population', *Journal of the American Statistical Association* **63**, 677–91.

Hill, B. 1970. 'Zipf's Law and Prior Distributions for the Composition of a Population', *Journal of the American Statistical Association* **65**, 1220–32.

Hill, B. 1979. 'Posterior Moments of the Number of Species in a Finite Population, and the Posterior Probability of Finding a New Species', *Journal of the American Statistical Association* **74**, 668–73.

Hill, B. 1988. 'Parametric Models for A_n: Splitting Processes and Mixtures', unpublished manuscript.

Hintikka, J. and Niiniluoto, I. 1980. 'An Axiomatic Foundation for the Logic of Inductive Generalization', in R. C. Jeffrey (ed.), *Studies in Inductive Logic and Probability*, Vol. 2, Berkeley and Los Angeles: University of California Press, pp. 157–82.

Hoppe, F. 1984. 'Polya-Like Urns and the Ewens Sampling Formula', *Journal of Mathematical Biology* **20**, 91–94.

Hoppe, F. 1987. 'The Sampling Theory of Neutral Alleles and an Urn Model in Population Genetics', *Journal of Mathematical Biology* **25**, 123–59.

Jeffrey, R. C. (ed.) 1980. *Studies in Inductive Logic and Probability*, Vol. 2, Berkeley and Los Angeles: University of California Press.

Johnson, William Ernest 1924. *Logic, Part III: The Logical Foundations of Science*, Cambridge, UK: Cambridge University Press.

Johnson, William Ernest 1932. 'Probability: the Deductive and Inductive Problems', *Mind* **41**, 409–23.

Kingman, J. F. C. 1975. 'Random Discrete Distributions', *Journal of the Royal Statistical Society* **B37**, 1–22.

Kingman, J. F. C. 1978a. 'Random Partitions in Population Genetics', *Proceedings of the Royal Society* **A361**, 1–20.

Kingman, J. F. C. 1978b. 'The Representation of Partition Structures', *Journal of the London Mathematical Society* **18**, 374–80.

Kingman, J. F. C. 1980. *The Mathematics of Genetic Diversity*, SIAM, Philadelphia.

Kuipers, T. A. F. 1973. 'A Generalization of Carnap's Inductive Logic', *Synthese* **25**, 334–36.

Kuipers, T. A. F. 1986. 'Some Estimates of the Optimum Inductive Method', *Erkenntnis* **24**, 37–46.

Laplace, P. S., Marquis de 1781. 'Mémoire sur les probabilités', *Mem. Acad. Sci. Paris 1778*, 227–32 (*Oeuvres complètes*, Vol. 9, pp. 383–485).

Quetelet, A. 1846. *Lettres à S.A.R. le Duc Régnant de Saxe-Cobourg et Gotha, sur la théorie des probabilités, appliquée aux sciences morales et politiques*, Brussels: Hayez.

Thisted, R. and Efron, B. 1987. 'Did Shakespeare Write a Newly-Discovered Poem?', *Biometrika* **74**, 445–55.

Zabell, S. L. 1982. 'W. E. Johnson's "Sufficientness" Postulate', *Annals of Statistics* **10**, 1091–99.

Zabell, S. L. 1988. 'Symmetry and its Discontents', in B. Skyrms and W. L. Harper (eds.), *Causation, Chance, and Credence*, Vol. 1, Dordrecht: Kluwer, pp. 155–90.

Zabell, S. L. 1989. 'The Rule of Succession', *Erkenntnis* **31**, 283–321.

11

The Continuum of Inductive
Methods Revisited

Let $X_1, X_2, X_3, \ldots$ denote a sequence of observations of a phenomenon (for example, the successive letters in an encrypted text, the successive species observed in a previously unexplored terrain, the success or failure of an experimental surgical procedure). In the classical Johnson-Carnap continuum of inductive methods (Johnson 1932, Carnap 1952), the outcomes that can occur are assumed to be of $T < \infty$ possible types or species that are known and equiprobable prior to observation. If, in a sample of n, there are n_1 outcomes of the first type, n_2 of the second, and so on, then (under appropriate conditions) the Johnson-Carnap continuum gives as the conditional epistemic probability of observing an outcome of the ith type on the $(n + 1)$-st trial the value

$$f(n_i, n) =: \frac{n_i + \alpha}{n + T\alpha} \ (\alpha > 0).$$

Note an important consequence of this: if $t < T$ species have been observed during the first n trials, then the probability of observing a new species on the next trial is a function of t and n,

$$g(t, n) =: 1 - \sum_{i=1}^{t} \frac{n_i + \alpha}{n + T\alpha} = \frac{T\alpha - t\alpha}{n + T\alpha};$$

by assumption, of course, $g(t, n) = 0$ for all $t \geq T$.

From its inception, the Johnson-Carnap continuum has been the subject of considerable controversy, and a number of its limitations have been pointed out by its critics; see, for example, the discussions in Howson and Urbach 1989 and Earman 1992. Among the most important of these is the failure of the continuum to permit the confirmation of universal generalizations and its assumption that the possible types that can arise are known in advance.[1]

In this essay I discuss a new three-parameter continuum of inductive methods, discovered by the statistician Pitman, which has a number of attractive features. First, it not only permits the confirmation of universal generalizations, but its mathematical derivation reveals this to be one of its essential elements. Second, it does not assume that the species to be observed are either known in advance or limited in number. Third, it interweaves two distinct continua: the observation of a new species both establishes a new category for confirmation *and* increases the likelihood of observing further new species down the road.

Of course, it is possible to achieve these desiderata in an ad hoc fashion: the confirmation of universal generalizations can certainly be achieved by assigning point masses to the initial probabilities in the de Finetti representation (Wrinch and Jeffreys 1919); an unlimited number of categories can be accommodated if one abandons the requirement that they be epistemically symmetric (see, e.g., Zabell 1982); and there have been several proposals in the literature regarding the use of rules of succession to predict the occurrence of new species (De Morgan 1845, 414–15; Kuipers 1973; Zabell 1992).

The compelling aspect of the new system discussed here is that all three of these features emerge as a natural consequence of a new postulate: as before, it is assumed that if a species has been observed n_i times out of n in the past, then the probability of observing that species again on the next trial is a function $f(n_i, n)$ of n_i and n alone, and that the probability of observing a new species is a function $g(t, n)$ of t (the number of species observed thus far) and n, but it is *not* assumed that $g(t, n) = 0$ for t greater than some prespecified value.

This essay is divided into four parts: in the first and second, the new continuum is explained, and some of its philosophical consequences explored; in the third, some prior literature is discussed; and in the fourth, the mathematical derivation of the continuum is given.

1. EXCHANGEABLE RANDOM PARTITIONS

The Johnson-Carnap continuum gives probabilities for exchangeable random *sequences*; the continuum discussed here gives probabilities for exchangeable random *partitions*.[2] In brief, if the different possible species are known in advance, it is possible to state the probability of seeing a particular sequence of individuals; if the different species are not known in advance, then it is only possible to state probabilities – *prior to observation* – for events framed in terms of the first species to be encountered, the second species to be encountered, and so on. (That is, one can state before the event "the species that occurs on the first trial will also occur on the third and fourth trials," but one

244

cannot state the event, "a giant panda will occur on the first, third, and fourth trials," unless one already knows that giant pandas exist.)

Thus, if a total of t different species are represented in a sequential sample of n individuals (observed at times $1, 2, \ldots, n$), and A_j is the set of times at which the jth species encountered is observed, then the sets $A_1, A_2, \ldots, A_t$ form a partition of the time set $\{1, 2, \ldots, n\}$; and it is to such partitions that probabilities are assigned. For example, suppose one encounters the sequence of transmitted symbols

QUOUSQUETANDEMABUTERECATALINAPATIENTIANOSTRA

There are $t = 26$ different possible letters that can occur in the sequence; there are a total of $n = 44$ letters observed in the text; the observed frequencies are $n_a = 7$, $n_t = 6$, $n_e = 5$, $n_i = n_n = n_u = 4$, $n_o = n_q = n_r = n_s = 2$, $n_b = n_c = n_d = n_l = n_m = n_p = 1$; and all other frequencies are zero.

Suppose, however, that a Romulan having no prior knowledge of our civilization encountered this sequence of symbols. It would not know in advance the 26 symbols in our alphabet. Thus, it notes that a total of $t = 16$ different symbols occur in the initial segment of length 44; that the first symbol encountered (the symbol "Q") occurred at positions 1 and 6 (that is, $A_1 = \{1, 6\}$); that the second symbol encountered (the symbol "U") occurred at positions 2, 4, 7, and 17 (that is, $A_2 = \{2, 4, 7, 17\}$), and so on. The sets A_1, $A_2, \ldots, A_{16}$ generate a partition $<A_1, A_2, \ldots, A_{16}>$ of the set $\{1, 2, \ldots, 44\}$. (See Table 1). The point is that – not knowing beforehand of the existence of our alphabet – a Romulan can hardly be expected to describe beforehand, let alone assign probabilities to, events such as the above 44-symbol sequence, but it is certainly possible for such a being both to describe and assign probabilities to the possible partitions (such as $<A_1, A_2, \ldots, A_{16}>$) that might arise from such a sequence.

In strictly mathematical terms, a *random partition* Π_n is a random object whose values are partitions π_n of a set $\{1, 2, \ldots, n\}$, and in the sampling of species problem it is precisely such random entities that one must consider (see Zabell 1992 for further discussion). In order to derive a continuum of inductive methods for such random partitions, some assumptions naturally have to be made concerning the underlying random structure governing their behavior. The four assumptions made here fall naturally into two classes or categories: one assumption is of a general nature, parallel to Johnson's *permutation postulate* for sequences (Johnson 1924); the other three assumptions are much more restrictive, parallel to Johnson's *sufficientness postulate* (Johnson 1932), and limit the possible random partitions that can arise to a three-parameter family. The general assumption that the random partitions

Table 11.1. *Delabeling*

In the first column, the S_i indicates the *i*th symbol or species to be observed in the example; the second column records this symbol; the third column, the subset of times when this species occurs; the fourth column, the size n_i of this subset (the number of times the *i*th species occurs); the fifth column, a_j, the number of species that occur *j* times (that is, the number of times the number *j* occurs in the preceding column). The resulting (unordered) partition of *n* is conveniently summarized as $1^6 2^4 3^1 4^2 5^1 6^1 7^0 8^1$.

S_1	Q	$\{1, 6\}$	$n_1 = 2$	$a_1 = 6$
S_2	U	$\{2, 4, 7, 17\}$	$n_2 = 4$	$a_2 = 4$
S_3	O	$\{3, 40\}$	$n_3 = 2$	$a_3 = 1$
S_4	S	$\{5, 41\}$	$n_4 = 2$	$a_4 = 2$
S_5	E	$\{8, 13, 19, 21, 34\}$	$n_5 = 5$	$a_5 = 1$
S_6	T	$\{9, 18, 24, 32, 36, 42\}$	$n_6 = 6$	$a_6 = 1$
S_7	A	$\{10, 15, 23, 25, 29, 31, 38, 44\}$	$n_7 = 8$	$a_7 = 0$
S_8	N	$\{11, 28, 35, 39\}$	$n_8 = 4$	$a_8 = 1$
S_9	D	$\{12\}$	$n_9 = 1$	–
S_{10}	M	$\{14\}$	$n_{10} = 1$	–
S_{11}	B	$\{16\}$	$n_{11} = 1$	–
S_{12}	R	$\{20, 43\}$	$n_{12} = 2$	–
S_{13}	C	$\{22\}$	$n_{13} = 1$	–
S_{14}	L	$\{26\}$	$n_{14} = 1$	–
S_{15}	I	$\{27, 33, 37\}$	$n_{15} = 3$	–
S_{16}	P	$\{30\}$	$n_{16} = 1$	–

CHECK: $44 = \sum_{j=1}^{44} j a_j = 1 \times 6 + 2 \times 4 + 3 \times 1 + 4 \times 2 + 5 \times 1 + 6 \times 1 + 8 \times 1$.

Π_n be *exchangeable* is discussed in the remainder of this section; the other three assumptions are discussed in the next.

Thus, let us consider the definition of an exchangeable random partition. Consider a partition $\pi = <A_1, A_2, \ldots, A_t>$ of the set $\{1, 2, \ldots, n\}$, and let $n_i =: n(A_i)$ denote the number of elements in the set A_i. Corresponding to π is the *frequency vector*

$$\mathbf{n} = \mathbf{n}(\pi) =: <n_1, n_2, \ldots, n_t>.$$

In turn, let a_j denote the number of frequencies n_i equal to j; then corresponding to the frequency vector is the *partition vector* (or "allelic partition")

$$\mathbf{a} = \mathbf{a}(\pi) =: <a_1, a_2, \ldots, a_n>.$$

(Because $n_j \leq n$, the number of components of the partition vector never can exceed n.) In Table 11.1 the various processes of "delabeling" a sequence, partitioning the time set $\{1, 2, \ldots, n\}$ into subsets A_j, and computing the species frequencies n_i and the components a_j of the partition vector $\mathbf{a}$ are illustrated for the example discussed above.

The random partition Π_n is said to be an *exchangeable random partition* (the concept is due to J. F. C. Kingman) if all partitions $<A_1, A_2, \ldots, A_t>$ having the same partition vector have the same probability; that is, if π_1 and π_2 are partitions of $(1, 2, \ldots, n\}$, then

$$\mathbf{a}(\pi_1) = \mathbf{a}(\pi_2) \Rightarrow P[\Pi_n = \pi_1] = P[\Pi_n = \pi_2].$$

In brief, the partition vector is a set of "sufficient statistics" for the random partition.[3] There is a sense in which this definition is natural; if one takes an exchangeable *sequence*, requires that its probabilities be category symmetric, and passes to the random partition it generates, then such partitions are exchangeable in the above sense. Note that the partition vector specifies an "unordered partition" of the number n; thus, an exchangeable random partition assigns the same probability to all partitions of the set $\{1, 2, \ldots, n\}$ that give rise to the same unordered partition of the number n.

Thus far we have considered one exchangeable random partition Π_n. Suppose we have an infinite sequence of them: $\Pi_1, \Pi_2, \Pi_3, \ldots$, such that Π_n is a partition of $\{1, 2, \ldots, n\}$ for each $n \geq 1$. There is a natural sense in which such a sequence of partitions $\{\Pi_n : n \geq 1\}$ is *consistent*: if $m < n$, then the random partition Π_n of $\{1, 2, \ldots, n\}$ induces a random partition of $\{1, 2, \ldots, m\}$; and one requires that for all $m < n$ that this induced random partition (denote it $\Pi_{m,n}$) coincides with the random partition Π_m (that is, one has $\Pi_{m,n} = \Pi_m$ for all $m < n$, $1 \leq m < n < \infty$). (To be precise, given a set T and a subset S of T, the partition $<A_1, A_2, \ldots, A_t>$ of T induces the partition $<A_1 \cap S, A_2 \cap S, \ldots, A_t \cap S>$ of S. The probability of the induced partition π of $\{1, 2, \ldots, m\}$ is then the probability of all partitions π^* of $\{1, 2, \ldots, n\}$ that induce π.) If $\{\Pi_n : n \geq 1\}$ is an infinite consistent sequence of exchangeable random partitions, then the random partition Π of the integers that $\{\Pi_n : n \geq 1\}$ gives rise to is also said to be exchangeable.

The simplest examples of infinite consistent sequences of exchangeable random partitions (or *partition structures*) are Kingman's *paintbox processes* (see, e.g., Aldous 1985, 87): one runs an infinite sequence of independent and identically distributed random variables $Z_1, Z_2, \ldots, Z_n, \ldots$, and then "delabels" the sequence (that is, passes to the partitions $\Pi_1, \Pi_2, \ldots, \Pi_n, \ldots$ of the time set that such a sequence gives rise to).[4]

There is a rich theory here: Just as the general infinite exchangeable sequence is formed from a mixture of independent and identically distributed sequences of random variables (this is the celebrated *de Finetti representation theorem*), the general infinite exchangeable random partition is a mixture of paintbox processes (this is the *Kingman representation theorem*). Indeed, every element of the classical theory of inductive inference for exchangeable random sequences has a counterpart in the theory of exchangeable random partitions (see Zabell 1992). But this powerful mathematical machinery is not needed here; just as in the classical Johnson-Carnap approach, it is possible in special cases to deduce directly from simple postulates the predictive probabilities of the partition structure characterized by those postulates; and this is the approach taken below.

This approach, although it has the twin merits of expository simplicity and philosophical clarity, reverses the actual process of historical discovery. The two-parameter family of partitions structures $\Pi_{\alpha,\theta}$ discussed in the next section were not discovered via their predictive probabilities. The Berkeley statistician Jim Pitman, working instead from the perspective of the Kingman representation, originally discovered them via their "residual allocation model" (RAM) characterization, derived their predictive probabilities from the RAM, and then suggested to me that it might be possible to characterize them by the form of their predictive probabilities (along the lines discussed in Zabell 1982). This paper states and proves a sharply formulated version of Pitman's conjecture.

It turns out that in the simplest such characterization it is necessary to add in a component corresponding to the confirmation of universal generalizations. The result is, therefore, a pleasant surprise: a simple three-parameter continuum of inductive methods that meets a fundamental objection to Carnap's original continuum for random sequences. The corresponding three-parameter family of partition structures already appears in Pitman's work: Corollary 3 in Pitman (1992b) characterizes the distribution of the "size-biased permutation" of the atoms in the Kingman representation of exactly such structures. The fact that this three-parameter family admits of two very different characterizations, both natural in their own right, is perhaps not without mathematical and philosophical significance.

In sum, the new continuum proposed here does *not* assume that the possible categories or species are known in advance. Successive observations are made, from time to time new species are encountered, and at each stage the number of outcomes so far noted in each category is recorded. At any given time n a total of t species have been encountered, and the number of outcomes that fall into each category is summarized by the partition vector $\mathbf{a} =: <a_1, a_2, \ldots, a_n>$, where a_1 is the number of species that appear once in the sample, a_2 is the number of species that appear twice in the sample, and so on.

In technical terms, it is assumed in the new continuum that *the probabilities that describe such a process give rise to an infinite exchangeable random partition.*

2. THE NEW CONTINUUM

Consider the problem of predicting the next outcome, given a sample of size n, for an infinite exchangeable random partition. The first assumption (1) made is:

$$P[\Pi_n = \pi_n] > 0 \text{ for all partitions } \pi_n \text{ of } \{1, 2, \ldots, n\}; \qquad (1)$$

that is, no particular species scenario is ruled out or deemed, a priori, to be impossible.

Let $Z_{n+1} \in S_i$ denote the event that the $(n+1)$st individual to be observed turns out to be a member of the i-th species to have already occurred. The second assumption (2) made is:

$$P[Z_{n+1} \in S_i | <n_1, n_2, \ldots, n_t>] = f(n_i, n), \quad 1 \le i \le t. \qquad (2)$$

That is, the *predictive probability* of observing the ith species on the next trial depends only on the number n_i of that species thus far observed, and the total sample size n. (Note the function $f(n_i, n)$ does not depend on the species i.)[5]

The third and final assumption (3) made is:

$$P[Z_{n+1} \in S_{t+1} | <n_1, n_2, \ldots, n_t>] = g(t, n). \qquad (3)$$

That is, the probability of observing a new species (since t species have been observed to date, this is necessarily the $(t+1)$st to be observed), is a function of the number of species thus far observed and the total sample size. It is a remarkable fact (proved in section 4) that if just these three conditions are imposed for all $n \ge 1$, then the functions $f(n_i, n)$ and $g(t, n)$ must be members of a three-dimensional continuum having parameters θ, α, and γ:

The Continuum of Inductive Methods for the Sampling of Species

Case 1. Suppose $n_i < n$ (and therefore $t > 1$; the universal generalization is *disconfirmed*). Then

$$f(n_i, n) = \frac{n_i - \alpha}{n + \theta}; \qquad g(t, n) = \frac{t\alpha + \theta}{n + \theta}.$$

Case 2. Suppose $n_i = n$ (and therefore $t = 1$; the universal generalization is confirmed). Then

$$f(n_i, n) = \frac{n - \alpha}{n + \theta} + c_n(\gamma) \qquad g(t, n) = \frac{\alpha + \theta}{n + \theta} - c_n(\gamma)$$

249

Increment due to confirmation of universal generalization:
Here

$$c_n(\gamma) = \frac{\gamma(\alpha+\theta)}{(n+\theta)\left[\gamma + (\alpha+\theta-\gamma)\prod_{j=1}^{n-1}\left(\frac{j-\alpha}{j+\theta}\right)\right]}.$$

The predictive probabilities in Case 1 are precisely the ones that Pitman (1992c) derived from the $\Pi_{\alpha,\theta}$ process. The numbers $c_n(\gamma)$ in Case 2 represent adjustments to these probabilities that arise when only one species is observed; in the present language, of seeing the partition $<A_1>$ consisting of the single set $A_1 = \{1, 2, 3, 4, \ldots, n\}$. The parameter γ is related to the prior probability ϵ that only one species is observed in an infinite sequence of trials; it is shown in the last section, "Derivation of the Continuum," that $\gamma = (\alpha+\theta)\epsilon$.[6]

If an infinite exchangeable random partition $\Pi = \{\Pi_1, \Pi_2, \Pi_3, \ldots\}$ satisfies Assumptions (1)–(3), then its predictive probabilities must have the above form for *some* α, θ, and γ. Not all values of α, θ, and γ are possible however. The question therefore arises as to which values of the parameters α, θ, and γ can actually be realized by an exchangeable random partition. (Of course, if such a random partition exists, it is necessarily unique because it can be generated by the given predictive probabilities). It is not difficult to prove that such exchangeable random partitions exist in precisely the following cases:

Range of possible parameter values in the continuum:

$$0 \leq \alpha < 1; \quad \theta > -\alpha; \quad \text{and} \quad 0 \leq \gamma < \alpha+\theta.$$

This corresponds to the new inductive continuum, discussed here.[7]

The subfamily of partitions $\Pi_{\alpha,\theta}$ that arise in the special case when $\gamma = 0$ are, as mentioned at the end of the previous section, the discovery of the Berkeley statistician Jim Pitman (1992a–d), who has extensively studied their properties; for this reason, it is referred to below as the *Pitman family* of infinite exchangeable random partitions.[8] (Other values of α and θ are possible if one either relaxes Assumption (1) or does not require the exchangeable random partition to be infinite; for simplicity, these cases are not discussed here.) The important class of partition structures that arise in the further special case $\alpha = 0$ had been discovered earlier by Warren Ewens (see Zabell 1992).

The two parameters α and θ can be interpreted as follows. The parameter θ is related to the a priori probability of observing a new species; the parameter α is the effect of subsequent observation on this likelihood. Given a new species, the information corresponding to the *first* observation of this species plays two roles and is divided into two parts. First, the observation is of a new species; thus, it gives us some reason to think that subsequent observations

250

will continue to give rise to new species, and it contributes the term $\frac{\alpha}{n+\theta}$ to $g(t, n)$. Second, the observation is of a particular species; thus the observation also gives us some reason to think that the species in question will continue to be observed in the future, and it contributes the term $\frac{1-\alpha}{n+\theta}$ to $f(n_i, n)$. In contrast, the parameter θ is related to the a priori likelihood of observing a new species, because no matter how many species have been observed, there is always a contribution of $\frac{\theta}{n+\theta}$ to $g(t, n)$.[9]

There is a simple urn model that describes the new continuum in the case $\gamma = 0$ and $\theta > 0$. Imagine an urn containing both colored and black balls, from which balls are both drawn and then replaced. Each time a colored ball is selected, it is put back into the urn, together with a new ball of the same color having a unit weight; each time a black ball is selected, it is put back into the urn together with two new balls, one black and having weight α ($0 \leq \alpha < 1$), and one of a new color, having weight $1 - \alpha$.

Initially the urn contains a single black ball (the *mutator*) having a weight of θ. Balls are then successively drawn from the urn; the probability of a particular ball being drawn at any stage is proportional to its selection weight. (The mutator is always selected on the first trial.) It is not difficult to see that the predictive probabilities at a given stage n are those of the Pitman $\Pi_{\alpha,\theta}$ process. If the special value $\alpha = 0$ is chosen, the resulting urn model reduces to that of the Hoppe urn (Hoppe 1984), and the predictive probabilities are those of the Ewens family $\Pi_\theta =: \Pi_{0,\theta}$.[10]

Remarks. In the Johnson-Carnap setting (prior known categories), symmetry among categories is purchased at the price of assuming that the total number of possible species $t < \infty$ (because $P[X_1 = x] = \frac{1}{t} > 0$, t must be finite). This does not happen in the case of the new continuum; it is possible for the number of species to be unbounded because no distinction is made between species at the time of the first observation. (Of course, *once* the first individual is observed, its species is then identified to the extent that subsequent individuals are classified as either LIKE or UNLIKE.)

In the Johnson-Carnap continuum, the confirmation of universal generalizations is not possible: if it were, then the predictive probability $f(0, n)$ for a known but unobserved category would depend on whether the observed number of species was greater than one. (That is, if $t = 1$ and $n_i = n$ for some cell i.) But in the new continuum, *there are no* "unobserved" categories; thus, the confirmation of universal generalizations becomes possible.

Given a sample of size n, the probability that an old species occurs on the next trial is $1 - \frac{t\alpha+\theta}{n+\theta} = \frac{n-t\alpha}{n+\theta}$; thus the conditional probability, given that an old species occurs, that it is of type i is $\frac{n_i-\alpha}{n-t\alpha}$. It is interesting to note that

251

this is of the same form as the Carnap continuum, but uses a negative value for the parameter α. Such predictive probabilities can arise in the classical setting in the case of a finite sequence (e.g., Zabell 1982) and also arise in Kuipers's work (see Kuipers 1978, chap. 6). Note also that in the special case $\alpha = 0$ the predictive probabilities reduce to Carnap's *straight rule*.

In the Pitman family $\mathbf{\Pi}_{\alpha,\theta}$ the number of observed species $t \to \infty$ as the sample size $n \to \infty$ (see my final remark in the last section of this essay). To avoid this, it is necessary to relax Assumption 1, that all cylinder sets have positive probability. To simplify the discussion, such questions are not examined in this paper.[11]

There are (at least) two very different ways in which both the classical Johnson-Carnap continuum and the new continuum proposed here may be viewed. One of these, initially that of Carnap (his later views were more complex), is that the probabilities in question are *objective*: that is, the three assumptions enunciated at the beginning of this section accurately capture the concept of ignorance regarding possible outcomes, and the continuum, therefore, gives the logical probabilities, credibilities (to use Russell's terminology), or rational degrees of belief regarding those possible outcomes. The other, polar extreme is to view the probabilities in question as *subjective* or personal; this is the view of the present author. In this case the three assumptions may be regarded as possible descriptors of our current epistemic state. If the three *qualitative* assumptions do in fact accurately describe our actual degrees of belief, then the force of the result is that our *quantitative* personal probabilities are uniquely determined up to three parameters.[12]

The characterization given in this section completes a parallelism in the theories of inductive inference for multinomial sampling and the sampling of species problem discussed in Zabell 1992 (see Table 11.2).

3. PRIOR PHILOSOPHICAL LITERATURE

The continuum of inductive methods discussed here has the advantage that it simultaneously meets two of the most important objections to the original Johnson-Carnap continuum: that the categories are not empirical in origin, and that universal generalizations cannot be confirmed. Both of these objections are in fact quite old, and in this section some of the past attempts and criticisms that have been made concerning these two points are discussed.

Richard Price and His Appendix

It is interesting to note that perhaps the first discussion concerning the origin of the categories used in the probabilistic analysis of inductive inference

Table 11.2. *Two Theories of Inductive Inference*

	Multinomial Sampling	Sampling of Species
Types	t types $(S_1, S_2, \ldots, S_t)$	initially unknown
Sample	random sequence $X = (X_1, X_2, \ldots, X_n)$	random partition $\mathbf{\Pi}$ of $\{1, 2, \ldots, n\}$
Sufficient statistics	sample frequencies $\mathbf{n} = (n_1, n_2, \ldots, n_t)$	allelic partition $\mathbf{a} = (a_1, a_2, \ldots, a_n)$
Exchangeability	$\mathbf{n}(x_1) = \mathbf{n}(x_2) \Rightarrow P[X = x_1] = P[X = x_2]$	$\mathbf{a}(\pi_1) = \mathbf{a}(\pi_2) \Rightarrow P[\mathbf{\Pi} = \pi_1] = P[\mathbf{\Pi} = \pi_2]$
Representation theorem	De Finetti representation theorem	Kingman representation theorem
Atomic constituents	i.i.d. sequences	paintbox processes
Canonical processes	Dirichlet priors	Pitman family $\mathbf{\Pi}_{\alpha,\theta}$
Urn model	Polya urn	Hoppe urn ($\alpha = 0$)
Sampling formula	Bose-Einstein statistics ($\alpha = 1$)	Ewens sampling formula ($\alpha = 0$)
Predictive probabilities	$\frac{n_i + \alpha}{n + t\alpha}$ ($\alpha > 0$)	$\frac{n_i - \alpha}{n + \theta}$, $\frac{t\alpha + \theta}{n + \theta}$ ($0 \leq \alpha < 1, \theta > -\alpha; t > 1$)
Characterization	$f(n_i, n)$; W. E. Johnson (1932)	RAM characterization; Pitman (1992b)
		$f(n_i, n)$, $g(t, n)$; see Section 4

goes back to Bayes's original essay (1764) or, more precisely, to an appendix to that essay penned by Bayes's friend, intellectual executor, and fellow dissenting Presbyterian clergyman, the Reverend Dr. Richard Price (1723–1791).

Bayes had considered "an event concerning the probability of which we absolutely know nothing antecedently to any trials made concerning it" (1764, 143). At the heart of Bayes's analysis is his famous (or infamous) postulate that in such cases all values of the probability p of such an event are equilikely. Given this assumption, it is easy to see that the chance that p falls between the limits a and b,[13] given the further information that the event has occurred n times in unfailing succession, is

$$P[a < p < b] = (n + 1) \int_a^b p^n dp = (n + 1) \left[\frac{b^{n+1} - a^{n+1}}{n + 1} \right]$$
$$= b^{n+1} - a^{n+1}.$$

Gillies (1987, 332) terms this *Price's rule of succession*, to distinguish it from the usual *Laplace rule of succession*: if there have been k successes in n trials, then the probability of a success on another trial is $\frac{k+1}{n+2}$. In particular the chance that the probability p lies between $\frac{1}{2}$ and 1 is $1 - (\frac{1}{2})^{n+1}$; in other words, the odds in favor of this are $2^{n+1} - 1$ to 1. Price's rule of succession is an immediate consequence of Bayes's results; his intellectual executor Price carried the analysis further in an appendix to Bayes's essay. Curiously, Price's appendix is often neglected in discussions of Bayes's essay, despite its great interest for students of inductive inference; two notable (and excellent) exceptions are Gillies 1987 and Earman 1992, chap. 1.

Price's analysis consists of several stages. In these the roles of the first and second observations play key roles. Price begins by considering the case of an uninterrupted string of successes, "a given number of experiments which are unopposed by contrary experiments." First, Price argues, prior to observation, all possible outcomes must have infinitesimal probability:

Suppose a solid or die of whose number of sides and constitution we know nothing; and that we are to judge of these from experiments made in throwing it.

In this case, it should be observed, that it would be in the highest degree improbable that the solid should, in the first trial, turn any one side which could be assigned beforehand; because it would be known that some side it must turn, and that there was an infinity of other sides, or sides otherwise marked, which it was equally likely that it should turn.

A little further on, Price adds:

I have made these observations chiefly because they are all strictly applicable to the events and appearances of nature. Antecedently to all experience, it would be improbable as infinite to one, that any particular event, beforehand imagined, should follow the application of any one natural object to another; because there would be an equal chance for any one of an infinity of other events. (Bayes 1764, append.)

There are already several interesting issues that arise in these passages. First, we have the resort to the "urn of nature": the argument is initially framed in terms of an objective chance mechanism (here the many-sided die), followed by the assertion that the analysis is "strictly applicable to the events and appearances of nature."[14] Note also Price's repeated emphasis that the events in question must be ones that can come to mind *prior* to observation: the side of the die is one which "could be assigned beforehand"; the event in nature is one which could be "beforehand imagined."

Such introspection reveals that there is an infinite spectrum of equipossible outcomes: the die might have an "infinity of other sides," each of which is "equally likely"; there is "an equal chance for any one of an infinity of other events." But how can an infinite number of such events be equally likely? Price avoids saying that such events have zero probability: they are instead "in the highest degree improbable" or "improbable as infinite to one." Perhaps he thought that if an event has zero probability, then it is impossible; perhaps he even thought of the probabilities at issue as being instead infinitesimals.[15]

Price argues that the first observation of a species or type has a special significance:

The first throw [of the solid or die] only shews that *it has* the side then thrown, without giving any reason to think that it has it any one number of times than any other. It will appear, therefore, that *after* the first throw and not before, we should be in the circumstances required by the conditions of the present problem, and that the whole effect of this throw would be to bring us into these circumstances. That is: the turning the side first thrown in any subsequent single trial would be an event about the probability or improbability of which we could form no judgment, and of which we should know no more than it lay somewhere between nothing and certainty. With the second trial our calculations must begin; and if in that trial the supposed solid turns again the same side, there will arise the probability of three to one that it has more of that sort of sides than of *all* others. (Ibid.)

Thus – according to Price – the first observation of an event results in a belief change of a very non-"Bayesian" type indeed! The observation of the first event transforms the status of its probability in future trials from the known – but infinitesimal – to the unknown but finite; the probability lies "somewhere between nothing and certainty."

This point of Price links up with a recurring issue in later debates on probability and induction. In the subjective or personalist system of Ramsey and de Finetti, the status of initial probabilities is straightforward: such probabilities summarize our present knowledge prior to the receipt of further information.[16] The subjective account does not tell us what these initial probabilities should be, nor does it provide a mechanism for arriving at them. It is just a theory of consistency, plain and simple. But if one hopes instead for more, for a theory of probability as a unique system of rational degrees of belief, then it is natural to demand the basis for the initial probabilities that are used. In a purely Bayesian framework, however, these must come from other, earlier initial probabilities; these must in turn come from other, still earlier initial probabilities, *und so weiter*; it is "turtles all the way down." The usual move to avoid such an infinite regress is to assume that at some stage one reaches a state of "total ignorance" and then pass to a uniform prior on a set of alternatives or parameters by appealing to the so-called principle of insufficient reason. But this is absurd: our ability to even *describe* an event in our language (or understand the meaning of a term used to denote an event) already implies knowledge – considerable knowledge – about that event.

The argument that Bayes employs instead to justify his choice of the uniform prior is much more subtle. His concern is "an event concerning the probability of which we absolutely know nothing antecedently to any trials made concerning it." That is, our ignorance pertains not to a knowledge of the circumstances of the event itself, but to its *probability*. It is sometimes thought that at this point Bayes then immediately passes to a uniform prior on the probability p. But, in fact (in an often overlooked scholium), he argues that in cases of absolute ignorance, if S_n denotes the number of times the event has thus far occurred in n trials, then

$$P[S_n = k] = \frac{1}{n+1} \quad \text{for } 0 \le k \le n.$$

This is in effect a precise quantitative translation of the informal qualitative formulation that we "absolutely know nothing antecedently to any trials made concerning" the probability of the event. It then follows as a direct mathematical consequence of this assumption (for all $n \ge 1$) that the prior distribution for the unknown probability p must be the uniform distribution (although Bayes did not himself so argue; see Zabell 1988).

Price takes Bayes's argument one step further. He asks: Just when are we in such a state of ignorance concerning the probability of an event? How can we pass from our sense impressions, a knowledge of prior events experienced (this happened and this did not), to an ignorance of their probabilities? Price argues that Bayes's formula is only applicable to types that have already been observed to occur at least once; and that the correct value of n to use in the formula is then *one less* than the total number of times that that type has thus far been observed to occur. (To check this, note that Price asserts the odds to be 3 to 1 after the same side turns up a *second* time; this corresponds to taking $n = 1$ in Price's rule of succession.)

From the vantage point of the new continuum, this corresponds to using the value $\alpha = 1$ in the continuum; that is, the *entire* weight of the first observation is given over to the prediction of further new species to be observed, and none is given to the prediction that the particular type observed will recur. There is, however, an obvious problem here. If $\alpha = 1$, then the probability of observing a second member of a species, given that one has already been observed, is

$$\frac{n_i - \alpha}{n + \theta} = \frac{1 - 1}{1 + \theta} = 0.$$

(As a result, the value $\alpha = 1$ partitions the set $\{1, 2, \ldots, n\}$ into the n singleton sets $\{1\}, \{2\}, \ldots, \{n\}$.) This difficulty does not arise in the new continuum proposed here, due to the constraint $\alpha < 1$. Price in effect circumvents such difficulties by forbidding the computation of probabilities at the first stage. Thus for Price it is only with "the second trial our calculations must begin"; and it is the observation of a *second* member of the species that permits calculation (for now $n_i - \alpha = 2 - 1 = 1$, and the predictive probabilities do not vanish).

Of course such an apparently ad hoc procedure requires justification. Price argues:

The first experiment supposed to be ever made on any natural object would only inform us of one event that may follow a particular change in the circumstances of those objects; but it would not suggest to us any ideas of uniformity in nature, or give us the least reason to apprehend that it was, in that instance or in any other, regular rather than irregular in its operations. (Ibid.)

This statement also has a natural interpretation within the context of exchangeable random partitions. The support of a paintbox process has both a discrete and a continuous component: The discrete component corresponds to the different species that occur (and recur) with positive probability; the continuous component corresponds to the *hapax legomena*, the species that occur once and then disappear, never to be seen again. In effect Price is saying that the first observation of a species tells us only that it lies in the support of the underlying process, but not whether it lies in its discrete component (and hence is "regular in its operations") or in its continuous component (and hence is "irregular").

Price illustrates the process of inductive inference in the case of natural phenomena by a curious hypothetical:

> Let us imagine to ourselves the case of a person just brought forth into this world, and left to collect from his observation of the order and course of events what powers and causes take place in it. The Sun would, probably, be the first object that would engage his attention; but after losing sight of it the first night he would be entirely ignorant whether he should ever see it again. He would therefore be in the condition of a person making a first experiment about an event entirely unknown to him. But let him see a second appearance or one *return* of the Sun, and an expectation would be raised in him of a second return. . . . But no finite number of returns would be sufficient to produce absolute or physical certainty. (Ibid.)

This is, in fact, a direct attack on Hume. To see the close relation, consider the corresponding passage from Hume's *Enquiry Concerning Human Understanding*:

> Suppose a person, though endowed with the strongest faculties of reason and reflection, to be brought on a sudden into this world; he would, indeed, immediately observe a continual succession of objects, and one event following another; but he would not be able to discover anything further. He would not, at first, by any reasoning, be able to reach the idea of cause and effect. (Hume 1748, 42; see also 27)

The image that Hume conjures up of a philosophical Adam first experiencing the sights and sounds of nature soon became a commonplace of the Enlightenment: It later appears in one form or another in Buffon's *Histoire naturelle de l'homme* of 1749, Diderot's *Lettre sur les sourds et muets* of 1751, Condillac's *Traité des sensations* of 1754, and Bonnet's *Essai de psychologie* of 1754 and *Essai analytique sur les facultés de l'âme* of 1760.[17] (Readers of Mary Shelley's *Frankenstein* [1818] will also recognize here the origin of the opening lines of part 2, chap. 3 of that book, when Frankenstein's creation first awakens to see the sun.)

Despite these later discussions (concerned primarily with issues in associationist psychology), it is clear that Price has Hume in mind: His discussion is a point-by-point attack on Hume's skeptical philosophical stance.[18] Hume denies that experience can give (immediate) knowledge of cause and effect; Price believes that the calculus of probabilities provides a tool that enables us to see how a person can "collect from his observation of the order and course of events" in the world, what "causes take place in it." Nor is this the only question at issue. In his *Treatise on Human Understanding*, Hume had written,

In common discourse we readily affirm, that many arguments from causation exceed probability, and may be receiv'd as a superior kind of evidence. One wou'd appear ridiculous, who wou'd say, that 'tis only probable the sun will rise to-morrow, or that all men must dye; tho' 'tis plain we have no further assurance of these facts, than what experience affords us. (Hume 1739, 124)[19]

Using Hume's own example of the rising of the sun, Price argues that in the case of uniform experience,

instead of proving that events will *always* happen agreeably to [uniform experience], there will be always reason against this conclusion. In other words, where the course of nature has been the most constant, we can have only reason to reckon upon a recurrency of events proportioned to the degree of this constancy; but we can have no reason for thinking that there are no causes in nature which will *ever* interfere with the operations of the causes from which this constancy is derived, or no circumstances of the world in which it will fail. (Bayes 1764, append.)

Thus Price argues that one can never achieve certitude regarding a *single* outcome on the basis of the finite experience at our disposal. (This should be distinguished from the assertion that it is unreasonable to confirm a universal generalization on the basis of a finite segment of experience.)

Augustus De Morgan

The one other classical student of the calculus of probabilities who appears to have considered the question of the origin of the categories used in inductive inference is the English mathematician Augustus De Morgan (1806–1871). (De Morgan was the leading enthusiast of Laplacean probability in England during the first half of the nineteenth century.) But while Price had, in effect, advocated the use of the parameter $\alpha = 1$, De Morgan introduced a different approach, corresponding to a choice of $\theta = 1$!

In a lengthy encyclopedia article that is in large part an exposition of Laplace's *Théorie analytique des probabilités* of 1812, De Morgan concludes his discussion of Laplace's rule of succession by noting:

There remains, however, an important case not yet considered; suppose that having obtained t sorts in n drawings, and t sorts only, we do not take it for granted that these are all the possible cases, but allow ourselves to imagine there may be sorts not yet come out. (1845, 414)[20]

In his little book *An Essay on Probabilities*, De Morgan gives a simple illustration of how he believes one can deal with this problem:

When it is known beforehand that either A or B *must* happen, and out of $m + n$ times A has happened n times, and B n times, then ... it is $m + 1$ to $n + 1$ that A will happen the next time. But suppose we have no reason, except what we gather from the observed event, to know that A or B must happen; that is, suppose C or D, or E, &c. might have happened: then the next event might be A or B, or a new species, of which it can be found that the respective probabilities are proportional to $m + 1, n + 1$, and 1; so that though the odds remain $m + 1$ to $n + 1$ for A rather than B, yet it is now $m + 1$ to $n + 2$ for A against either B or the other event. (De Morgan 1838, 66)

De Morgan's prescription can be understood in terms of a Hoppe urn model in which initially there are three balls, one labeled "A," one "B," and one "black," the *mutator*. Balls are then selected from the urn as follows: balls labeled "A" or "B" (or any other letter or symbol) are replaced together with another of the same label; if the mutator is selected, then it is replaced by a ball labeled by a new letter or symbol not yet encountered. The resulting exchangeable random partition corresponds to a conditional Pitman process (after the observation of one "A" and one "B"), having parameters $\alpha = 0$ and $\theta = 1$.[21]

Recent Literature

It is an important historical footnote that Carnap thought that the sampling of species problem could be dealt with by introducing a predicate relation R: IS THE SAME SPECIES AS. But correctly recognizing the considerable increase in complexity this would introduce into the problem, Carnap did not pursue this idea further. (The information in this paragraph is due to Richard Jeffrey.)

There have been few attempts in the recent philosophical literature to deal with such problems since Carnap; this is not entirely surprising if one accepts the basic thesis of this paper, that the machinery of exchangeable random partitions is crucial in coming to grips with them. Hintikka 1966 and Hintikka and Niiniluoto 1980 consider cases where the predictive probabilities depend, not just on the sample size n and number of instantiations k, but also the number of species observed. Their results assume, however, that the total spectrum of possible species is both known and finite; for further information, see Kuipers 1978, which contains a careful and detailed analysis of these systems. (It would be interesting to derive results parallel to theirs for exchangeable random partitions.)

Kuipers himself dealt with the problem in an early paper (1973). Kuipers's proposal interweaves two continua. One is binomial: on each trial, a new species does or does not occur with probability $\frac{t+\lambda}{n+2\lambda}$; the other is multinomial: conditional on a new species not occurring, if k instances of a species have already been observed, then the probability that the species recurs is $\frac{k+\mu}{n+t\mu}$. (Note that the machinery of random partitions is implicit in Kuipers's proposal: a random partition, rather than a random sequence, is generated because the character of the new species to appear is not stated; it is just a new species.) Unfortunately, the random partitions so generated are not exchangeable: the probability of seeing an old species and then a new species does not equal the probability of seeing a new species and then an old one.

Kuipers (1978, chap. 6, sec. 10) also considered the possibility of extending Hintikka's results to the case of an infinite number of alternatives, and discusses what he describes as a "reformulation of an H-system." This turns out to be nothing other than the "delabeling" process described earlier.[22] Because delabeling is equivalent to passing to the underlying partition generated by a sequence, Kuipers's approach is equivalent to viewing matters from the perspective of random partitions. It is of particular interest to note that Kuipers proposes as the predictive probability of a new species, given n observations to date *and supposing that an infinite number of species are ultimately observed,* $\frac{\lambda_\infty}{n+\lambda_\infty}$, $0 \leq \lambda_\infty < \infty$. These conditional probabilities correspond in the present continuum to the special case $\alpha = 0$ (and $\theta = \lambda_\infty$), that is, the *Ewens subfamily.*

Confirmation of Universal Generalizations

The failure of Carnap's original continuum to confirm universal generalizations is too well known to require any but the briefest discussion here. (It is perhaps worth noting, however, that the issue itself is actually quite old [see Zabell 1989, 308–09 for a number of references prior to Popper and going back to the nineteenth century].) Barker (1957, 84–90) summarizes a number of the early objections. Popper's primary assault (1959, 363–77) was effectively rebutted by Howson 1973; other later criticisms include Essler 1975. There are a number of good discussions that can serve as entries into this literature; these include Kuipers (1978, 96–99) and Earman (1992, 87–95).

4. DERIVATION OF THE CONTINUUM

Let us begin by using Johnson's original argument (Johnson 1932; Zabell 1982; Costantini and Galavotti 1987) and see where it leads, *assuming (as we do throughout this section) that the three postulates of the second section, "The New Continuum," hold.* The first step in the argument is to prove that for each $n \geq 1, f(n_i, n)$ is linear in n_i. This turns out to be nearly true here too. Recall that $\delta_n(k) = 0$ for $1 \leq k < n$, and $\delta_n(n) = 1$.

Lemma 1. *For each $n \geq 1$, there exist constants a_n, b_n, and c_n such that*

$$f(k, n) = a_n + b_n k + c_n \delta_n(k)$$

for $1 \leq k \leq n$.

Proof. If $n = 1, 2$, or 3, it is immediate that the desired equation holds for a suitable choice of coefficients a_n, b_n, and c_n (since the number of constraints is at most three). Thus, we may assume $n \geq 4$. Let $f(n_i, n) = f(n_i)$ and $g(t, n) = g(t)$. Choose $c_n =: f(n) - f(n-1)$. It suffices to prove

$$f(n_i + 1) - f(n_i) = f(n_i) - f(n_i - 1), \quad 1 < n_i < n - 1.$$

Suppose that at stage n there are t species, and that the frequency count is $(n_1, n_2, \ldots, n_t)$. Consider the ith species, and suppose that $1 < n_i < n - 1$. Because $n_i < n$, there exists at least one other species j; suppose that $n_j > 1$ (this is possible because $n_i < n - 1$). Then one has:

$$f(n_i) + f(n_j) + \sum_{k \neq i, j} f(n_k) + g(t) = 1.$$

Because $n_i > 1$, one can remove an individual from species i without extinguishing the species, and use it to create a new species; one then has for the resulting partition:

$$f(n_i - 1) + f(n_j) + \sum_{k \neq i, j} f(n_k) + f(1) + g(t + 1) = 1.$$

Equating the two and subtracting then gives:

$$f(n_i) - f(n_i - 1) = g(t + 1) - g(t) + f(1).$$

Likewise, by taking one element from j and creating a new species, we get

$$f(n_j) - f(n_j - 1) = g(t + 1) - g(t) + f(1).$$

Finally, take one from j, and put it into i; then

$$f(n_i + 1) + f(n_j - 1) + \sum_{k \neq i, j} f(n_k) + g(t) = 1,$$

and

$$f(n_i + 1) - f(n_i) = f(n_j) - f(n_j - 1),$$

hence

$$f(n_i + 1) - f(n_i) = f(n_i) - f(n_i - 1).$$

This concludes the proof. □

Next, let us consider the effect of the sample size n. Suppose $b_n = 0$; then $f(n_i, n)$ is independent of n_i (except in the case of a universal generalization). In the Johnson-Carnap setting, one separates out such possibilities by showing that if b_n vanishes at a single stage n, then it vanishes at *all* stages. In the present setting, however, this never happens.

Lemma 2. *For all $n \geq 3$, $b_n \neq 0$.*

Proof. Consider a given sequence of observations up to stage n. Because the random partition is exchangeable, the observation of an old species once more at stage $n + 1$ and then a new species at stage $n + 2$ generates a partition having the same probability as the partition that arises from observing the new species at stage $n + 1$ and the old at stage $n + 2$. Suppose that $b_n = 0$. If $n_i < n$, then $t \geq 2$ (at stage n) and

$$(1 - ta_n)(a_{n+1} + b_{n+1}n_i) = a_n(1 - ta_{n+1} - b_{n+1}(n + 1)),$$

hence

$$(1 - ta_n)b_{n+1}n_i = a_n(1 - b_{n+1}(n + 1)) - a_{n+1};$$

thus $(1 - ta_n)b_{n+1}n_i$ is constant (as a function of n_i) for n_i in the range $1 \leq n_i < n$. It thus follows that if $n \geq 3$ (so that both $n_i = 1$ and $n_i = 2$ are possible), then either $1 - ta_n = 0$ or $b_{n+1} = 0$. If $1 - ta_n = 0$, then $a_n = t^{-1}$. But if $n \geq 3$, then both $t = 2$ and $t = 3$ are possible. Thus $a_n = t^{-1}$ is impossible (since a_n is a constant), hence $b_{n+1} = 0$. Thus: if $b_n = 0$ and $n \geq 3$, then $b_{n+1} = 0$. (Note that $b_n = 0 \rightarrow b_{n+1} = 0$ need not hold for $n = 1$ or 2, because b_n is not uniquely determined in these two cases.)

It immediately follows that if $b_n = 0$ for $n = n_0 \geq 3$, then $b_n = 0$ for all $n \geq n_0$. But it is easy to see that this cannot happen. For if $b_n = 0$, then $f(n_i, n) = a_n$ and $g(t, n) = 1 - t_n a_n$ for $t \geq 2$. Then, arguing as before, we see that

$$(1 - t_n a_n)a_{n+1} = a_n(1 - t_n a_{n+1}) \rightarrow a_{n+1} = a_n.$$

Thus if b_n vanishes from some point on, then a_n is constant from this point on. Let a denote the resulting common value of a_n, $n \geq n_0$. Since $a > 0$, it follows that $na > 1$ for n large. But this is impossible, because $1 - na = g(n, n) > 0$. Thus $b_n \neq 0$ for all $n \geq 3$. □

263

Because b_n does not vanish, one can normalize a_n and c_n relative to it; thus let

$$\alpha_n =: -\frac{a_n}{b_n}, \; \gamma_n =: \frac{c_n}{b_n}, \quad \text{and} \quad \theta_n =: \frac{g(t_n, n)}{b_n} + t_n \frac{a_n}{b_n}.$$

Since

$$t a_n + b_n n + g(t, n) = \sum_{i=1}^{t} (a_n + b_n n_i) + g(t, n) = 1 \text{ for } t \geq 2,$$

$$b_n^{-1} = n + t \frac{a_n}{b_n} + \frac{g(t, n)}{b_n} = n + \theta_n,$$

hence

$$f(n_i, n) = a_n + b_n n_i + c_n \delta_n(n_i) = \frac{n_i + \frac{a_n}{b_n} + \frac{c_n}{b_n} \delta_n(n_i)}{b_n^{-1}}$$

$$= \frac{n_i - \alpha_n + \gamma_n \delta_n(n_i)}{n + \theta_n},$$

$$g(t, n) = 1 - t a_n - b_n n - c_n \delta_1(t) = \frac{t \alpha_n + \theta_n - \gamma_n \delta_1(t)}{n + \theta_n}.$$

Such normalization is also possible even in the special cases $n = 1$ and $n = 2$ not covered by Lemma 2: if $n = 1$, the probability of observing at the second trial the same species as on the first is $a_1 + b_1 + c_1$; if $n = 2$ the probabilities of observing a species, given it has been observed once or twice are $a_2 + b_2$ and $a_2 + 2b_2 + c_1$; and it is clear that in both cases we are free to choose b_1 and b_2 so that neither vanishes.

Lemma 3. *For $n \geq 3$, α_n does not depend on n and θ_n does not depend on either n or t.*

Proof. Step 1: $\alpha_n = \alpha_{n+1}$. If $n_i < n$, then there exist at least two categories: label these i and j. Consider two possibilities: you observe (1) a member of species i at time $n + 1$, j at time $n + 2$; (2) a member of species j at time $n + 1$, and i at time $n + 2$. Because the random partition is exchangeable, the conditional probabilities of these two possibilities are the same, hence

$$\left(\frac{n_i - \alpha_n}{n + \theta_n} \right) \left(\frac{n_j - \alpha_{n+1}}{n + 1 + \theta_{n+1}} \right) = \left(\frac{n_j - \alpha_n}{n + \theta_n} \right) \left(\frac{n_i - \alpha_{n+1}}{n + 1 + \theta_{n+1}} \right),$$

hence $n_i(\alpha_n - \alpha_{n+1}) = n_j(\alpha_n - \alpha_{n+1})$. Because $n \geq 3$, we can choose $n_i = 1$, $n_j = 2$; and it thus follows that $\alpha_n = \alpha_{n+1}$.

Step 2: $\theta_n = \theta_{n+1}$. Next, consider the two possibilities; you observe (1) a member of a new species at time $n + 1$, and a member of the old species i at time $n + 2$; (2) a member of the old species i at time $n + 1$ and a member of a new species at time $n + 2$. Equating the conditional probabilities of these two events gives us:

$$\left(\frac{t\alpha_n + \theta_n}{n + \theta_n} \right) \left(\frac{n_i - \alpha_{n+1}}{n + 1 + \theta_{n+1}} \right) = \left(\frac{n_i - \alpha_n}{n + \theta_n} \right) \left(\frac{t\alpha_{n+1} + \theta_{n+1}}{n + 1 + \theta_{n+1}} \right).$$

Because we already know that $\alpha_n = \alpha_{n+1}$, it follows that $\theta_n = \theta_{n+1}$. □

The two predictive probabilities $f(n_i, n)$ and $g(t, n)$ are therefore seen to be of the form

$$f(n_i, n) = \frac{n_i + \alpha + \gamma_n \delta_n(n_i)}{n + \theta}, \quad g(t, n) = \frac{t\alpha + \theta - \gamma_n \delta_1(t)}{n + \theta},$$

for some α and θ, and all $n \geq 3$. It is not difficult to see, however, that these two formulas continue to hold in the special cases $n = 1$ and 2 (using the same values for α and θ), provided appropriate choices are made for γ_1 and γ_2. This is trivial if $n = 1$; one just chooses an appropriate value for γ_1 given the already determined values of α and θ. For the case $n = 2$, note that one can choose $\alpha_2 = \alpha$ because there are three degrees of freedom (α_2, θ_2, and γ_2), but only two constraints involving predictive probabilities (the values of $f(1, 2)$ and $f(2, 2)$). Thus it remains to show that $\theta_2 = \theta$; but this follows using the same argument in Step 2 of Lemma 3 (using $n = t = 2$, together with the observation that $1 - \alpha$ does not vanish because $f(1, 2)$ does not, by assumption, vanish).

Note, however, that the formulas derived state only that the desired conditional probabilities must have the given form for *some* θ and α (and sequence γ_n); they do not assert that for each possible pair $<\alpha, \theta>$ an exchangeable random partition exists that satisfies our assumptions for all such values. Indeed, it is not hard to see that certain constraints are essential:

Lemma 4. *The parameters α and θ satisfy $0 \leq \alpha < 1$ and $\theta > -\alpha$.*

Proof. For any fixed value of θ, $n + \theta > 0$ for all n sufficiently large. Because all possible finite sequences have positive probability, $0 < g(t, n) < 1$, hence

$$0 < \frac{t\alpha + \theta}{n + \theta} < 1 \Rightarrow -\frac{\theta}{t} < \alpha < \frac{n}{t}.$$

265

Letting $n \to \infty$ and taking $t = n$, it follows that $0 \le \alpha < 1$. Because $0 < f(1, 1) < 1$,

$$0 < \frac{1 - \alpha}{1 + \theta} < 1;$$

but $1 - \alpha > 0$, hence $1 + \theta > 0$, hence $\theta > -\alpha$. □

On the other hand, exchangeable random partitions do exist for all possible $<\alpha, \theta>$ pairs in the ranges given in Lemma 4 (see Pitman 1992). It thus remains to identify the constant γ_n. In order to do this, however, a technical result is required.

Lemma 5 (Basic recurrence relation).

$$\gamma_{n+1} = \frac{\gamma_n(n + \theta)}{(n - \alpha + \gamma_n)}.$$

Proof. By the same argument as in Lemma 3, observe that

$$\left(\frac{n - \alpha + \gamma_n}{n + \theta}\right) \left(\frac{\theta + \alpha - \gamma_{n+1}}{n + 1 + \theta}\right) = \left(\frac{\theta + \alpha - \gamma_n}{n + \theta}\right) \left(\frac{n + 1 - \alpha}{n + 1 + \theta}\right);$$

canceling denominators and some simplification then gives the result. □

Let $\gamma =: \gamma_1$, and for $n \ge 1$, let $\Pi_n =: (\frac{1-\alpha}{1+\theta})(\frac{2-\alpha}{2+\theta}) \cdots (\frac{n-\alpha}{n+\theta})$ and $d_n =: \gamma + (\alpha + \theta + \gamma)\Pi_{n-1}$ (by convention, $\Pi_0 = 1$).

Lemma 6. *For all $n \ge 1$, $\gamma_n = \frac{\gamma(\alpha+\theta)}{d_n}$.*

Proof. It suffices to prove that (1) $d_n \ne 0$ and (2) $\gamma_n d_n = \gamma(\alpha + \theta)$ for all $n \ge 1$. The proof is by induction.

Note first that $d_1 = \gamma + (\alpha + \theta - \gamma) = \alpha + \theta > 0$, and $\gamma_1 d_1 = \gamma(\alpha + \theta)$; thus the two assertions hold for $n = 1$. Next, suppose that $d_n \ne 0$ and $\gamma_n d_n = \gamma(\alpha + \theta)$ for a given value of $n \ge 1$. Then

$$
\begin{aligned}
d_{n+1} &=: \gamma + (\alpha + \theta - \gamma) \prod_n \\
&= \gamma + (\alpha + \theta - \gamma) \left(\frac{n - \alpha}{n + \theta}\right) \prod_{n-1} \\
&= \frac{(n - \alpha)\left(\gamma + (\alpha + \theta - \gamma) \prod_{n-1}\right) + \gamma(\alpha + \theta)}{n + \theta}
\end{aligned}
$$

$$= \frac{(n-\alpha)d_n + \gamma_n d_n}{n+\theta}$$

$$= \frac{(n-\alpha+\gamma_n)d_n}{n+\theta}.$$

Thus $d_{n+1} = \frac{(n-\alpha+\gamma_n)d_n}{n+\theta}$.

But then $d_{n+1} \neq 0$ immediately follows from the inductive hypothesis (because $n - \alpha + \gamma_n > 0$ and $n + \theta > 0$ for all $n \geq 1$); and

$$\gamma_{n+1}d_{n+1} = \frac{\gamma_n(n+\theta)}{(n-\alpha+\gamma_n)}d_{n+1} = \gamma_n d_n = \gamma(\alpha+\theta)$$

(by the fundamental recursion formula, the preceding formula, and the inductive hypothesis). □

Lemma 7. *The parameter γ satisfies the inequalities $0 \leq \gamma < \alpha + \theta$.*

Proof. Suppose $\gamma < 0$; since $n + \theta > 0$ and $n - \alpha + \gamma_n > 0$ for all n, it follows from the fundamental recursion formula that $\gamma_n < 0$ for all n. But $\lim_{n\to\infty} \Pi_n = 0$, hence

$$\gamma_n = \frac{\gamma(\alpha+\theta)}{\gamma + (\alpha+\theta-\gamma)\Pi_{n-1}} \to \alpha + \theta > 0,$$

which is impossible. Thus $\gamma \geq 0$. The inequality $\gamma < \alpha + \theta$ follows from the inequalities $1 + \theta > 0$ and

$$\frac{1-\alpha+\gamma}{1+\theta} = f(1,1) < 1.$$

□

The basic result stated in the second section, "The New Continuum," now follows immediately from Lemmas 1–7. The following theorem restates this in terms of a mixture of two partitions, and summarizes the primary technical contribution of this paper regarding the characterization of exchangeable random partitions.

Theorem 1. *Let $\Pi = \Pi_1, \Pi_2, \ldots, \Pi_n, \ldots$ be an infinite consistent sequence of exchangeable random partitions. If the sequence satisfies the three Assumptions (1), (2), and (3) for all $n \geq 1$, then there exist three parameters ϵ, α, and θ ($0 \leq \epsilon < 1, 0 \leq \alpha < 1$, and $\theta > -\alpha$), such that*

$$f(n_i, n) = (1 - \epsilon_n)\left(\frac{n_i - \alpha}{n+\theta}\right) + \epsilon_n \delta_n(n_i)$$

and

$$g(t, n) = (1 - \epsilon_n) \left(\frac{t\alpha + \theta}{n + \theta} \right) - \epsilon_n \delta_1(t),$$

where

$$\epsilon_n =: \frac{\epsilon}{\epsilon + (1 - \epsilon) \prod_{j=1}^{n-1} \frac{j-\alpha}{j+\theta}}$$

is the posterior probability of the partition Π_∞ (all observations are of the same species), given that the first n observations are of the same species, an initial probability of ϵ in favor of Π_∞ and $1 - \epsilon$ in favor of the Pitman alternative $\Pi_{\alpha,\theta}$.

Proof. Note that if $z > x$, then

$$\frac{x + y}{z} = (1 - r)x + r \leftrightarrow r = \frac{y}{z - x}.$$

Thus letting $x = n - \alpha$, $y = \gamma_n$, $z = n + \theta$, and $r = \frac{\gamma_n}{\alpha + \theta} =: \epsilon_n$ gives

$$\frac{n - \alpha + \gamma_n}{n + \theta} = (1 - \epsilon_n) \left(\frac{n - \alpha}{n + \theta} \right) = \epsilon_n.$$

Let $\epsilon =: \frac{\gamma}{\alpha + \theta}$; it then follows from Lemma 6 that ϵ_n has the stated form. Because $0 \leq \gamma < \alpha + \theta$ (Lemma 7), it follows that $0 \leq \epsilon < 1$. That ϵ_n is the stated posterior probability is an immediate consequence of Bayes' theorem. □

Remarks. 1. Suppose further that $g(t, n) = g(n)$ for all $n \geq 1$; that is, that the probability of seeing a new species depends only on the sample size n, and *not* the number of species observed. Then necessarily $\epsilon = \alpha = 0$, and one has a characterization of the Ewens subfamily $\Pi_{0,\theta}$.[23] To be precise:

Corollary. *Let Π be an infinite exchangeable random partition, and let Π_n denote the exchangeable random partition induced by Π on the set $\{1, 2, \ldots, n\}$. Suppose that for each $n \geq 1$, (1) $P[\Pi_n = \pi_n] > 0$ for all partitions π_n of $\{1, 2, \ldots, n\}$; (2) the conditional probability of observing an old species i at time $n + 1$, given Π_n, the past history up to time n, depends only on n and n_i, the number of times that the species i has occurred in the past (but not on i or n_j for $j \neq i$); and (3) the conditional probability of observing a new species at time $n + 1$, given Π_n, depends only on n. Then the random partition Π is a member of the Ewens family for some value $\theta > 0$; that is, $\Pi = \Pi_{0,\theta}$ for some $\theta > 0$.*

268

2. Consider the Pitman process $\Pi_{\alpha,\theta}$. If E_n is the event that a novel species is observed on the nth trial, and t_n is the number of distinct species observed as of that trial, then

$$\sum_{n=1}^{\infty} P[E_{n+1}|t_1, t_2, \ldots, t_n] = \sum_{n=1}^{\infty} \frac{t_n \alpha + \theta}{n + \theta} \geq \sum_{n=1}^{\infty} \frac{\theta}{n + \theta} = \infty;$$

it then follows from the extended Borel-Cantelli lemma (see, for example, Breiman 1968, 96, Corollary 5.29) that

$$P[E_n \text{ occurs infinitely often}]$$

$$= P\left[\omega: \sum_{n=1}^{\infty} P[E_{n+1}|t_1, t_2, \ldots, t_n] = \infty\right] = 1.$$

Thus, the total number of species observed in an infinite number of observations of the Pitman $\Pi_{\alpha,\theta}$ is almost surely infinite. (In fact, Pitman [1992c] shows that the number of species t_n grows almost surely as the power n^α: the random limit $Z =: \lim_{n\to\infty} t_n/n^\alpha$ exists almost surely and has a distribution that depends on θ.)

ACKNOWLEDGMENT

I thank Jim Pitman for drawing my attention to the $\Pi_{\alpha,\theta}$ family of random partitions; his conjecture that some form of the Johnson theorem should apply to it led to the present essay. Theo Kuipers was also very generous in providing information regarding his 1973 paper. Persi Diaconis, Warren Ewens, Theo Kuipers, and Jim Pitman made helpful comments on a first draft. It is a particular pleasure to acknowledge the hospitality of the Instituto di Statistica of the Università degli Studi di Genova, and that of its director, Domenico Costantini, during a visit to Genova in April 1992, when my initial research began.

NOTES

1. The confirmation of a universal generalization means that if only one species is observed, then an increased positive probability is assigned to the possibility that only one species exists.
2. Both the concept of an exchangeable random partition, and the closely allied concept of *partition structure*, are due to the English mathematician J. F. C. Kingman; see Kingman 1980, Aldous 1985, and Zabell 1992 for further information and references.

3. An early example of the use of the partition vector in the sampling of species problem can be found in letters of R. A. Fisher to his Cambridge colleague Sir Harold Jeffreys (see Bennett 1990, 151, 156–57, 160).

   During the second world war the English mathematician and logician Alan Mathison Turing (1912–1954) recognized the importance of such "frequencies of frequencies" and used them to break key German military codes at Bletchley Park (see Good 1965, chap. 7, and 1979). Turing's statistical interests in such problems are less surprising than they might at first seem: his 1935 undergraduate King's College fellowship dissertation proved a version of the Lindeberg central limit theorem, and this experience led him in later years to be on the alert for the potential statistical aspects of a problem (see Zabell 1995).

4. Strictly speaking, the term *partition structure* refers to the consistent sequence of random partition vectors generated by a consistent sequence of exchangeable random partitions.

5. How could it? In the Johnson-Carnap continuum, the fact that $f(n_i, n)$ does not depend on i is an *assumption*, but here it is a *consequence* of the framework; the ith species is not known to exist prior to sampling!

6. Thus with probability ϵ, all animals are of the same species as the first animal, and with probability $1 - \epsilon$, the predictive probabilities are $\frac{n_i - \alpha}{n + \theta}$. Theorem 1 in the final section of this essay, "Derivation of the Continuum," states the continuum in the alternative format of a mixture.

7. The two cases $\alpha = 1$ and $\gamma = \alpha + \theta$ are excluded because of Assumption (1): the case $\alpha = 1$ corresponds to the random partition where each species occurs only once; the case $\gamma = \alpha + \theta$ to the random partition where only one species occurs.

8. As noted at the end of the first section of this essay, Pitman has also investigated the more general family discussed here: Corollary 3 in Pitman (1992b) characterizes the distribution of the "size-biased permutation" of the atoms in the Kingman representation of exactly such partition structures. The ϵ in Theorem 1 of this essay's last section corresponds to Pitman's $P(P_1 = 1)$.

9. The role and interpretation of the α and θ parameters become much more complex at the level of the corresponding partition structures: for $\alpha > 0$ fixed, the laws of $\Pi_{\alpha,\theta}$ are mutually absolutely continuous as θ varies, but for θ fixed, the $\Pi_{\alpha,\theta}$ are mutually singular as α varies (see Pitman 1992c).

10. Such urns are the "delabeled" versions of urn models for Dirichlet processes that first appear in Blackwell and MacQueen 1973.

11. Thus, the classical Johnson-Carnap continuum for multinomial sampling and the continuum of inductive methods for the sampling of species discussed here represent two extremes, neither entirely credible. In one case (Johnson-Carnap), it is assumed that because of (but it is, in fact, in spite of) our supposed ignorance all possible categories are a priori known and equiprobable. In the other case (the sampling of species, considered here), it is assumed that one is ignorant of *all* categories in advance but, in spite of this, knows that an infinite number of them must occur over time. A more realistic continuum for the sampling of species would eliminate assumption (1). I hope to return to this question in joint work with Jim Pitman.

12. This in essence was Johnson's viewpoint: In his (posthumous) 1932 paper in *Mind*, he wrote:

> The postulate adopted in a controversial kind of theorem cannot be generalized to cover all sorts of working problems; so it is the logician's business, having once formulated a specific postulate, to indicate very carefully the factual and epistemic conditions under which it has practical value. (1932, 418–19)

13. In Bayes's terminology, probabilities pertain to events, chances to probabilities of events.

14. The urn of nature, in this sense, goes back to James Bernoulli's *Ars conjectandi*; the cogency of the analogy was a primary target for the later critics of inverse probabilities (see, e.g., Zabell 1989, 302–03).

15. Price states in a footnote, "There can, I suppose, be no reason for observing that on this subject unity is always made to stand for certainty, and $\frac{1}{2}$ for an even chance." But if unity stands for certainty, then presumably zero stands for impossibility. For a modern attempt to interpret a related species of inverse probabilities in terms of infinitesimals, see Sobel 1987.

16. For a discussion of Ramsey's system, as set forth in his 1926 essay, "Truth and Probability," see Zabell 1991.

17. For Buffon's philosophical Adam and Condillac's criticism of him, see Fellows and Milliken (1972, 125–31); for discussion of Condillac's alternative, the "statue-man," and the cited work of Diderot and Bonnet, see Knight (1968, chap. 4). These discussions do not cite Hume as a precursor; and it seems unlikely, given the proximity in dates, that Buffon, the earliest of them, drew the image directly from Hume a mere year after the appearance of the *Enquiry*. Perhaps there is a common intellectual ancestor at work here, but I have not as yet been able to find one. All of these French narratives reflect interests in associationist psychology; for the more general link between associationism and probability, see Daston (1988, chap. 4). These fanciful narratives reflected a more general Enlightenment fascination with persons reared in the wild or initially deprived of certain sensory abilities (see Gay 1969, 174–76).

18. Gillies 1987 presents an able and convincing argument of the case (see also Daston 1988, 264–67, 326–30). There is some evidence that Bayes himself intended his essay as an answer to Hume (see Zabell 1989, 290–93).

19. Hume's interpolation of "proofs" as third species of reasoning, intermediate between that of "knowledge" and "probabilities" and consisting of nondemonstrative arguments that are "entirely free of doubt and uncertainty," is similar to Cardinal Newman's concept of "assent" in his *Grammar of Assent* (1870).

20. I have modified De Morgan's notation to conform with mine. Although much of De Morgan's article is closely based on Laplace's book, I have not been able to find either a paper or book of Laplace in which this point is made. It appears to be original with De Morgan.

21. Strictly speaking, the urn model is not specified by De Morgan. Indeed, his discussion elsewhere suggests that he thought the appropriate denominator to use in the predictive probabilities to be $n + t + 1$, where n is the (nonrandom) sample size

and t the (random) number of species observed, rather than t the number of species known to exist a priori.

22. See Kuipers (1978, chap. 6, sec, 10). In Kuipers's system the predictive probabilities for old species can depend on the number of species observed. The infinite system is derived by first delabeling a finite system, and then passing to the limit as the number of species increases to infinity. It would be of considerable interest to have a direct axiomatic derivation of this system in the infinite case.

23. For other characterizations of the Ewens subfamily, see Kingman (1980, 38) and Donnelly (1986, 279–81).

REFERENCES

Aldous, D. J. 1985. "Exchangeability and related topics." In P. L. Hennequin, ed., *École d'Été de Probabilités de Saint-Flour 1983, Lecture Notes in Mathematics* 1117:1–198.

Barker, S. F. 1957. *Induction and Hypothesis*. Ithaca, N.Y.: Cornell University Press.

Bayes, Rev. T. 1764. "An essay towards solving a problem in the doctrine of chances." *Philosophical Transactions of the Royal Society of London* 53:370–418. Reprinted in E. S. Pearson and M. G. Kendall, (eds.), *Studies in the History of Statistics and Probability*, vol. 1. London: Charles Griffin, pp. 134–53. Page references are to this edition.

Bennett, J. H., ed. 1990. *Statistical Inference and Analysis: Selected Correspondence of R. A. Fisher*. Oxford, UK: Clarendon Press.

Blackwell, D., and J. B. MacQueen. 1973. "Ferguson distributions via Polya urn schemes." *Annals of Statistics* 1:353–55.

Breiman, L. 1968. *Probability*. New York: Addison-Wesley.

Carnap, R. 1952. *The Continuum of Inductive Methods*. Chicago: University of Chicago Press.

Costantini, D., and M. C. Galavotti. 1987. "Johnson e l'interpretazione degli enunciati probabilistici." In R. Simili, ed., *L'Epistemologia di Cambridge 1850–1950*, Società Editrice il Mulino, Bologna, pp. 245–62.

Daston, L. 1988. *Classical Probability in the Enlightenment*. Princeton, NJ: Princeton University Press.

De Morgan, A. 1838. *An Essay on Probabilities, and Their Application to Life Contingencies and Insurance Offices*. London: Longman, Orme, Brown, Green, & Longmans.

De Morgan, A. 1845. "Theory of probabilities." In *Encyclopedia Metropolitana*, vol. 2: *Pure Mathematics*. London: Longman et al.

Donnelly, P. 1986. "Partition structures, Polya urns, the Ewens sampling formula, and the ages of alleles." *Theoretical Population Biology* 30:271–88.

Earman, J. 1992. *Bayes or Bust? A Critical Examination of Bayesian Confirmation Theory*. Cambridge, Mass.: MIT Press.

Essler, W. K. 1975. "Hintikka vs. Carnap." In J. Hintikka, ed., *Rudolph Carnap, Logical Empiricist*, Dordrecht, Holland: D. Reidel Publishing Co.

Fellows, O. E., and S. F. Milliken. 1972. *Buffon*, New York: Twayne Publishers.

Gay, P. 1969. *The Enlightenment: An Interpretation*, vol. 2: *The Science of Freedom*. New York: W. W. Norton.

Gillies, D. 1987. "Was Bayes a Bayesian?" *Historia Mathematica* 14:325–46.

Good, I. J. 1965. *The Estimation of Probabilities: An Essay on Modern Bayesian Methods*. Cambridge, MA: MIT Press.

Good, I. J. 1979. "A. M. Turing's statistical work in World War II." *Biometrika* 66, 393–96.

Hintikka, J. 1966. "A two-dimensional continuum of inductive methods." In J. Hintikka and P. Suppes, (eds.), *Aspects of Inductive Logic*. Amsterdam: North-Holland, pp. 113–32.

Hintikka, J., and I. Niiniluoto. 1980. "An axiomatic foundation for the logic of inductive generalization." In R. C. Jeffrey, ed., *Studies in Inductive Logic and Probability*, vol. 2. Berkeley, Calif.: University of California Press, pp. 157–81.

Hoppe, F. 1984. "Polya-like urns and the Ewens sampling formula." *Journal of Mathematical Biology* 20:91–94.

Howson, C. 1973. "Must the logical probability of laws be zero?" *British Journal for Philosophy of Science* 24:153–63.

Howson, C., and P. Urbach. 1989. *Scientific Reasoning: The Bayesian Approach*. La Salle, III.: Open Court Press.

Hume, D. [1739] 1978. *A Treatise of Human Nature*. In the L. A. Selbe-Bigge text, revised by P. H. Nidditch. Oxford: Clarendon Press.

Hume, D. [1748] 1975. *An Enquiry Concerning Human Understanding*. In the L. A. Selbe-Bigge text, revised by P. H. Nidditch. Oxford: Clarendon Press.

Johnson, W. E. 1932. "Probability: the deductive and inductive problems." *Mind* 41:409–23.

Kingman, J. F. C. 1980. *The Mathematics of Genetic Diversity*. Philadelphia: SIAM.

Knight, I. F. 1968. *The Geometric Spirit*. New Haven and London: Yale University Press.

Kuipers, T. A. F. [1973] 1975. "A generalization of Carnap's inductive logic." *Synthese* 25:334–36. Reprinted in J. Hintikka, ed., *Rudolph Carnap, Logical Empiricist*. Dordrecht: D. Reidel Publishing Co.

Kuipers, T. A. F. 1978. *Studies in Inductive Probability and Rational Expectation*. Dordrecht: D. Reidel Publishing Company.

Newman, Cardinal J. H. 1870. *An Essay in Aid of a Grammar of Assent*, fifth ed. 1885. London: Longman & Green's.

Pitman, J. 1992a. "Partially exchangeable random partitions." Technical Report 343, Department of Statistics, University of California, Berkeley. Revised version to appear in *Probability Theory and Related Fields* 102(1995):145–58.

Pitman, J. 1992b. "Random discrete distributions invariant under size-biased permutation." Technical Report 344, Department of Statistics, University of California, Berkeley. To appear in *Advances in Applied Probability* 28(1996): 525–39.

Pitman, J. 1992c. "The two-parameter generalization of Ewens' random partition structure." Technical Report 345, Department of Statistics, University of California, Berkeley.

Pitman, J. 1992d. "Random partitions derived from excursions of Brownian motion and Bessel processes." Technical Report 346, Department of Statistics, University of California, Berkeley.

Popper, K. [1959] 1968. *The Logic of Scientific Discovery*. New York: Basic Books. 2nd English ed., New York: Harper & Row.

Shelley, M. [1818] 1992. *Frankenstein*, 2nd ed. 1831. Reprinted, London: Penguin Books.

Sobel, J. H. 1987. "On the evidence of testimony for miracles: A Bayesian interpretation of David Hume's analysis." *Philosophical Quarterly* 37:166–86.

Wrinch, D., and H. Jeffreys. 1919. "On certain aspects of the theory of probability." *Philosophical Magazine* 38:715–31.

Zabell, S. L. 1982. "W. E. Johnson's 'sufficientness' postulate." *Annals of Statistics* 10:1091–99.

Zabell, S. L. 1988. "Symmetry and its discontents." In *Causation, Chance, and Credence*, vol. 1, W. L. Harper and B. Skyrms, eds., Dordrecht: Kluwer, pp. 155–90.

Zabell, S. L. 1989. "The rule of succession." *Erkenntnis* 31:283–321.

Zabell, S. L. 1991. "Ramsey, truth, and probability." *Theoria* 57:211–38.

Zabell, S. L. 1992. "Predicting the unpredictable." *Synthese* 90:205–32.

Zabell, S. L. 1995. "Alan Turing and the central limit theorem." *American Mathematical Monthly* 102: 483–94.

Index

Gregory, D. F., 120
Gruder, O., 107

Hadamard's theorem, 205
Hall, Philip, 200
hapax legomena, 257
Hardy, Godfrey Harold, 209–210
Hartley, David, 45–46, 67
Hausdorff theorem, 7
Hilbert, David, 200
Hintikka, J., 92, 229, 260
Hoppe, Fred, 219, 222, 237, 239, 251
Hume, David, 18–19, 52, 68
   Enquiries, 46
   On Miracles, 79
   problem of induction, 3, 5–6
   Treatise, 47, 74

inductive behavior, 132–133
inductive probabilities, 11
infinite limiting frequencies, 4
inverse probability from 1880 to 1930,
   152–155
   textbook treatments, 155

Jaynes, Edwin T., 28
Jeffrey, Richard C., 130
Jeffreys, Sir Harold, 29–30, 33, 67, 90, 93,
   156, 187
   cordial relations with Fisher, 195
Jevons, William Stanley, 51–53, 55
Johnson, N. L., 107, 112
Johnson, William Ernest, 12, 31, 33, 123–124,
   239
   combination postulate, 8–10, 57, 224
   Keynes's debt to, 125–126, 137
   permutation postulate, 56–57, 84, 245
   sufficientness postulate, 10–11, 57–58,
     84–93, 225–226, 245

Kahneman, Daniel and Tversky, Amos, 28
Kemeny, John G., 93
Keynes, John Maynard, 29–30, 33, 54, 60, 93,
   123–126, 155, 200
   dispute with Pearson, 125, 137
   Treatise on Probability, 129–130, 137
   My Early Beliefs, 137
   recantation, 131
Kingman, J. F. C., 219, 238, 247
   representation theorem, 232–234,
     247–248
Kolmogorov, A. N., 210

Kraitchik, Maurice, 33
Von Kries, Johannes, 54, 129
Kuipers, Theo, 92, 239, 261

Lacroix, Sylvestre-Francois, 68
Lagrange and duration of play, 135
Laplace, Pierre Simon Marquis de, 19–22, 32,
   46, 52
   and example of the rising of the sun, 47–50,
     54, 69, 74
   citation practices, 78–80
   least squares, 135
   rule of succession, 68, 74, 219–220
Laudan, L., 51
Leibniz, G., 13, 14
Leucippus, 13
Lévy, Paul, 202–203, 205
Lindeberg, Jarl Waldemar, 200
   central limit theorem, 202
   condition, 202, 206–207
logic, material versus formal or conceptualist
   view, 122–123
London Bills of Mortality, 18

M-estimators, 135
Maskell, E. J., 165–166
mathematical probability, birth of, 17
mean absolute deviation, monotonicity of,
   110–111
Mill, John Stuart, 32, 79, 153
Mills ratio, 108
Milner-Barry, P. S., 211, 212
Mitchell, John, 19
Monmort, Pierre Rémond de, 103
   Correspondence with Bernoulli about game
     of "Le Her", 157
Moore, G. E.,
   influence on Bloomsbury, 137

Newman, John Henry Cardinal, 47
Newman, M. H. A., 200
Neyman, Jerzy, 26–27, 142, 154, 157, 187
Niiniluoto, I., 92, 229, 260

Oresme, Nicole, 16–17
Ostrogradskii, M. V., 41, 67
Owen, G. E. L., 13, 15
ou mallon, 15–16

paintbox process, 233, 247, 257–258
paradox
   two-envelope, 30